Carpenters and Builders Library Volume Three

Layouts, Foundations, Framing

by John E. Ball

Revised and Edited by Tom Philbin

★★★

The Bobbs-Merrill Co., Inc.
Indianapolis/New York

FIFTH EDITION
FIRST PRINTING—1982

Copyright © 1965, 1970, 1976 and 1978 by Howard W. Sams & Co., Inc.
Copyright © 1982 by The Bobbs-Merrill Co., Inc.
Printed in the United States of America

Library of Congress Cataloging in Publication Data

Ball, John E.
    Layouts, foundations, framing.

    (Carpenters and builders library / by John E. Ball ; v. 3)
    Includes index.
    1. Building layout. 2. Foundations. 3. Framing (Building) I. Philbin, Tom,
1934– . II. Title. III. Series: Ball, John E. Carpenters and builders library ; v. 3.
TH385.B34 1982   694   82–1339
ISBN 0–672–23367–3   AACR2

Published by The Bobbs-Merrill Company, Inc.
Indianapolis/New York

Foreword

This volume, the third in a complete library for the carpenter, do-it-yourselfer, and builder, gives special emphasis to the construction of the house and other buildings. The book starts at ground level, as it were, discussing the way buildings are laid out. Then it moves on to a discussion of the building's foundation—including the new, all-weather wooden foundation—and gets into the art of form making (and an art it is, requiring a special breed of carpenter).

Concrete block construction is explained and then the true anatomy of the home is explored, detailing everything the carpenter and craftsman need to know about building a house—or part of one. There is also much here that is valuable for the student of buildings in general.

While rooted in the basics of house construction, the book also discusses other areas, from porches and patios to stoves, that have become important in our energy-conscious society. So, too, there is an in-depth discussion of insulation—batts, blankets, rigid, fill—whatever has passed the test of time and the demands of the energy crisis.

Today, the carpenter must also know about solar energy, and this is discussed in Chapter 15. If present trends continue, the carpenter will be called on to do many solar installations.

Cosmetic features of house construction have not been excluded from this volume. Among other things, there is a step-by-step presentation on how to install paneling, which is the most important development in wall coverings in twenty-five years.

All in all, here is a volume that should prove valuable to many readers.

Contents

CHAPTER 1

Laying Out

The term laying out here means the process of locating and fixing the reference lines that define the position of the foundation and outside walls of a building.

SELECTION OF SITE

Preliminary to laying out (sometimes called staking out) it is important that the exact location of the building be properly selected. In this examination, it may be wise to dig a number of small, deep holes at various points, extending to a depth a little below the bottom of the basement.

If the holes extend down to its level, the ground water, which is sometimes present near the surface of the earth, will appear in the bottoms of the holes. This water will nearly always stand at the same level in all the holes.

If possible, a house site should be located so that the bottom of the basement is above the level of the ground water. This may mean locating the building at some elevated part of the lot or reducing the depth of excavation. The availability of storm and sanitary sewers, and their depth, should have been previously investigated. The distance of the building from the curb is usually stipulated in city building ordinances, but this, too, should be known.

STAKING OUT

After the approximate location has been selected, the next step is to lay out the building lines. The position of all corners of

Fig. 1. One way of laying out is with a hundred foot tape. Metal tape is standard, but this new fiberglass one works well and cleans easily. Courtesy of Stanley.

the building must be marked in some way so that when the excavation is begun, workers will know the exact boundaries of the basement walls. There are a couple of methods of laying out building lines:

1. With surveyor's instrument.
2. By method of diagonals.

The Lines

Several lines must be located at some time during construction, and they should be carefully distinguished. They are:

1. The line of excavation, which is the outside line.
2. The face line of the basement wall inside the excavation line, and in the case of masonry building.
3. The ashlar line, which indicates the outside of the brick or stone walls.

In a wooden structure, only the two outside lines need to be located, and often only the line of the excavation is determined at the outset.

Laying Out with Transit Instruments

A transit is, of course, an instrument of precision, and the work of laying out is more accurate than with other methods. In Fig. 2, let *ABCD* be a building already erected; at a distance from this (at right angle), building *GHJK* will be erected. Level up the instrument at point *E*, making *A* and *E* the distance the new building will be from points *A* and *B*. Make points *B* and *F* the same length as points *A* and *E*. At this point, drive a stake in the ground at point *G*, making points *F* and *G* the required distance between the two buildings. Point *H* will be on the same line as point *G*, making the distance between the two points as required.

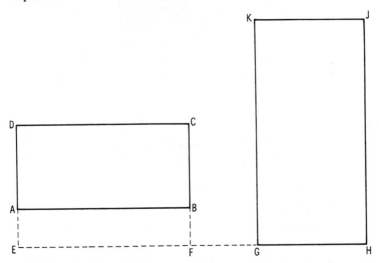

Fig. 2. Diagram illustrating method of laying out with transit.

Place the transit over point *G*, and level it up. Focus the transit telescope on point *E* or *F* and lock into position. Turn the horizontal circle on the transit until one of the zeros exactly coincides with the vernier zero. Loosen the clamp screw and turn the telescope and vernier 90 degrees. This will locate point

K, which will be at the desired distance from point *G*. For detailed operation of the transit, see the manufacturer's instructions or information in Volume 2 of this series. The level may be used in setting floor timbers, in aligning posts, and in locating drains.

Fig. 3. Transit, used by builders, contractors, and others for setting grades, batter boards, and various earth excavations.

Method of Diagonals

All that is needed in this method is a line, stakes, and a steel tape measure. Here, the right angle between the lines at the corners of a rectangular building is found by calculating the length of the diagonal which forms the hypotenuse of a right-angle triangle. By applying the following rule, the length of the diagonal (hypotenuse) is found.

Rule —The length of the hypotenuse of a right-angle triangle is equal to the square root of the sum of the squares of each leg.

Thus, in a right-angle triangle ABC, of which AC is the hypotenuse,

$$AC = \sqrt{AB^2 + BC^2} \dots\dots\dots\dots\dots\dots\dots\dots\dots\dots\dots(1)$$

Suppose, in Fig. 4, $ABCD$ represents the sides of a building to be constructed, and it is required to lay out these lines to the dimensions given. Substitute the values given in equation (1), thus,

$$AC = \sqrt{30^2 + 40^2} = \sqrt{900 + 1600} = \sqrt{2500} = 50$$

To lay out the rectangle of Fig. 4, first locate the 40-foot line AB with stake pins. Attach the line for the second side to B, and measure off on this line the distance BC (30 feet), point C being indicated by a knot. This distance must be accurately measured with the line at the same tension as in A and B.

With the end of a steel tape fastened to stake pin A, adjust the position of the tape and line BC until the 50-foot division on the tape coincides with point C on the line. ABC will then be a right angle and point C will be properly located.

The lines for the other two sides of the rectangle are laid out in a similar manner. After getting the positions for the corner

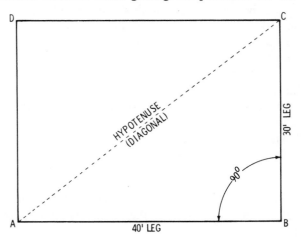

Fig. 4. Diagram illustrating how to find the length of the diagonal in laying out lines of a rectangular building by the method of diagonals.

stake pins, erect batter boards and permanent lines. A simple procedure may be used in laying out the foundations for a small rectangular building. Be sure that the opposite sides are equal and then measure *both* diagonals. No matter what this distance may be, they will be equal if the building is square. No calculations are necessary, and the method is precise.

POINTS ON LAYING OUT

For ordinary residence work, a surveyor or the city engineer is employed to locate the lot lines. Once these lines are established, the builder is able to locate the building lines by measurement.

A properly prepared set of plans will show both the contour of the ground on which the building is to be erected and the new grade line after the building is done. A convenient way of determining old grade lines and establishing new ones is by means of a transit, or with a Y level and a rod. Both instruments work on the same principle in grade work. As a rule, a masonry contractor has his own Y level and uses it freely as the wall is constructed, especially where levels are to be maintained as the courses of material are placed.

In locating the grade of the earth about a building, stakes are driven into the ground at frequent intervals and the amount of "fill" indicated by the heights of these stakes. Grade levels are usually established after the builders have finished, except that the mason will have the grade indicated for him where the wall above the grade is to be finished differently from the wall below grade. When a Y level is not available, a 12- or 14-foot straightedge and a common carpenter's level may be used, with stakes being driven to "lengthen" the level.

SUMMARY

The term "laying out" means the process of locating a fixed reference line that will indicate the position of the foundation and walls of a building.

A problem sometimes encountered is ground water. It is sometimes present near the surface of the earth and will appear in the bottoms of test holes, generally at the same level. If possible, a house should be located so that the bottom of the basement floor is above the level of the ground water.

After the location of the house has been selected, the next step is to lay out or stake out the building lines. The position of all corners of the house must be marked so that workmen will know the exact boundaries of the basement walls.

There are several ways to lay out a building site. Two of these are with a surveyor's instrument and with diagonal measurements. When laying out a site, several lines must be located at some time during construction. These lines are the line of excavation, which is the outside line; the face line of the basement wall inside the excavation line; and in the case of a masonry building, the ashlar line, which indicates the outside of the brick or stone wall.

REVIEW QUESTIONS

1. What is ground water?
2. Name two methods used in laying out a building site.
3. What is the difference between laying out and staking out?
4. What is the line of excavation?
5. What is the ashlar line?

CHAPTER 2

Foundations

The foundation is the part of a building that supports the load of the superstructure. As generally understood, the term includes all walls, piers, columns, pilasters, or other supports below the first-floor framing.

There are three general forms of foundation: spread foundations, pile foundations and wood foundations (Fig. 1). Spread foundations are the most popular type used. They receive the weight of the superstructure and distribute the weight to a stable soil base by means of individual footings. Pile foundations, on the other hand, transmit the weight of the superstructure through a weak soil to a more stable base. Of relatively recent vintage is the all-weather all-wood foundation, which is made of plywood soaked with preservatives.

There are three basic types of spread foundations: the slab-on-grade, the crawl space, and the basement or full, as shown in Fig. 2. Each foundation system is popular in certain geographic areas. The slab-on-grade is popular in the South and Southwest;

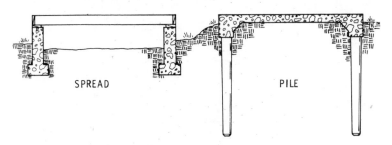

SPREAD   PILE

Fig. 1. General forms for foundations.

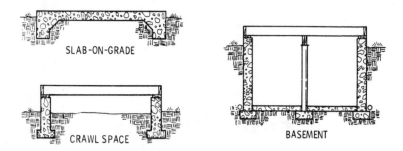

Fig. 2. Three types of spread foundations.

the crawl space is popular throughout the nation; and the basement is the most popular in the northern states.

SLAB-ON-GRADE

There are three basic types of slab-on-grade. The most popular is where the footing and slab are combined to form one integral unit. Another type has the slab supported by the foundation wall, and there is a type where the slab is independent of the footing and foundation wall (Fig. 3).

The procedure for constructing a slab-on-grade would be:

1. *Clear the site*. In most cases no excavation is needed, but

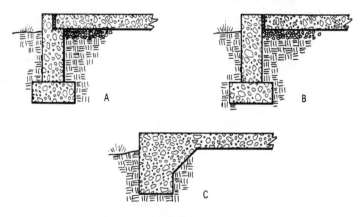

Fig. 3. Three types of slab-on-grade foundations.

some fill dirt may be needed. A tractor or bulldozer is usually used to remove the necessary brush and trees. It can also be used to spread the necessary fill.

2. *Lay out the foundation.* This is usually done with batter board and strings. When the batter boards are attached to the stakes, the lowest batter board should be 8 inches above grade.

3. *Place and brace the form boards.* The form boards are usually 2 × 12s, 2 × 6s, or 2 × 4s, and are aligned with the builders string. To keep the forms in proper position they are braced with 2 × 4s. One 2 × 4 is placed adjacent to the form board and another is driven at an angle three feet from the form board. Between the two 2 × 4s a "kicker" is placed to tie the two together. These braces are placed around the perimeter of the building, 4 feet on center (Fig. 4).

4. *Additional fill is brought in.* The fill should be free of debris or organic matter and should be screeded to within 8 inches of the top of the forms. The fill should then be well tamped.

5. *Footings are dug.* The footings should be a minimum of 12 inches wide and should extend 6 inches into undisturbed soil. The footings should also extend at least one foot below the frost line.

6. *Place the base course.* The base course is usually wash gravel or crushed stone and is placed 4 inches thick. The base course acts as a capillary stop for any moisture that might rise through the soil.

Fig. 4. Form construction.

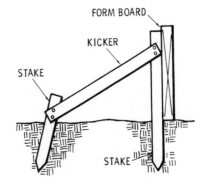

FORM BOARD

KICKER

STAKE

STAKE

7. *Place the vapor barrier.* The vapor barrier is a sheet of .006 polyethylene and acts as a secondary barrier against moisture penetration.
8. *Reinforce the slab.* In most cases, the slab is reinforced with 6 × 6 No. 10 gauge wire mesh. To ensure that the wire mesh is properly embedded, it is propped up or pulled up during the concrete pour.
9. *Reinforce the footings.* The footings can be reinforced with three or four deformed metal bars 18 to 20 feet in length. The rods should not terminate at a corner, but should be bent to project around it. At an intersection of two rods, there should be an 18-inch overlap.

Once the forms are set and the slab bed completed, concrete is brought in and placed in position. The concrete should be placed in small piles and as near to its final location as possible. Small areas of concrete should be worked—in working large areas the water will supersede the concrete, causing inferior concrete. Once the concrete has been placed in the forms, it should be worked (poked and tamped) around the reinforcing

Fig. 5. Hand tamping or "jitterbugging" concrete to place large aggregate below the surface. The other man (bent over) is screeding with a long 2 × 4 to proper grade.

bars and into the corners of the forms. If the concrete is not properly worked, air pockets or honeycombs may appear.

After the concrete has been placed, it must be struck or screeded to the proper grade. A long straightedge is usually used in the process. It is moved back and forth in a sawlike motion until the concrete is level with the forms. To place the large aggregate below the surface, the concrete is hand tamped or "jitterbugged" (Fig. 5). A darby is used immediately after the jitterbug and is also used to embed the large aggregate (Fig. 6). To produce a "round" on the edge of the concrete slab, an edger is used. The "round" keeps the concrete from chipping off and it increases the aesthetic appeal of the slab (Fig. 7). After the water sheen has left the surface of the slab, it is floated. Floating is used to remove imperfections and to compact the surface of the concrete. For a smooth and dense surface the concrete is then troweled. It can be troweled with a steel hand trowel or it can be troweled with a power trowel (Fig. 8).

Once the concrete has been finished, it should be cured. There are three ways that the slab might be cured: (1) burlap coverings, (2) sprinkling, (3) and ponding. Regardless of the technique used, the slab should be kept moist at all times.

Fig. 6. The darby being used after the jitterbug process.

Fig. 7. Using an edger to round off the edges.

Fig. 8. Using a power trowel.

A CRAWL SPACE

A crawl space foundation system can be constructed of an independent footing and concrete block foundation wall, or the footing and foundation wall can be constructed of concrete.

The footing should be constructed of concrete and should be placed below the frost line. The projection of the footing past the foundation wall should equal ½ the thickness of the foundation wall (Fig. 9). The thickness of the footing should equal the width of the foundation wall. There are two basic ways to form a footing for a crawl space. One of the most popular ways is to dig a footing trench and place the concrete in the trench. To maintain the proper elevation, grade stakes are placed in the trench. The other technique is to use form boards. If form boards are used they should be properly erected and braced. In some cases, additional strength may be needed, and reinforcement added.

The most convenient way to obtain concrete for a crawl space foundation or footing, or for that matter, any job where a fair amount of material is required, is to have it delivered by truck. The mix will be perfect, and it will be poured exactly where you are ready for it with a minimum of effort on your part. Of course, this method is more expensive than mixing it yourself. If you mix it yourself, you can rent mixing equipment, such as a power mixer. A good strong concrete mix is three bags of sand to every bag of cement, and enough water to keep the mix

Fig. 9. Footing construction.

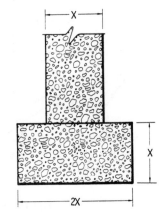

workable. Or you can use four bags of concrete sand—sand with rocks in it—to every bag of cement. Forms are removed after the concrete has hardened. Before laying the concrete masonry, the top of the footings should be swept clean of dirt or loose material.

Regardless of whether the foundation wall is constructed of placed concrete or concrete blocks, the top should be a minimum of 18 inches above grade. This allows for proper ventilation, repair work, and visual inspection.

BASEMENT CONSTRUCTION

Foundation walls, in basement construction, should be built with the utmost care and craftsmanship because they are under great pressure from water in the ground.

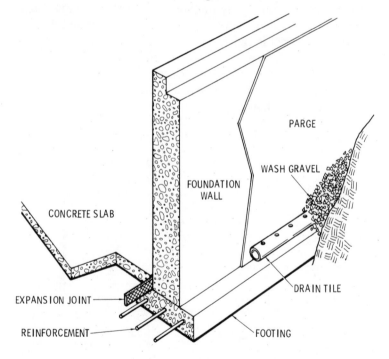

PARGE

WASH GRAVEL

FOUNDATION WALL

CONCRETE SLAB

DRAIN TILE

EXPANSION JOINT

REINFORCEMENT

FOOTING

Fig. 10. Basement construction.

To properly damp-proof the basement if such a situation exists, a 4-inch drain tile can be placed at the base of the foundation wall. The drain tile can be laid with open joints or it can have small openings along the top. The tile should be placed in a bed of wash gravel or crushed stone and should drain into a dry stream bed or storm sewer. Outside the wall should then be parged with masonry cement, mopped with hot asphalt, or covered with polyethylene. These techniques will keep moisture from seeping through the foundation. For further protection, all surface water should be directed away from the foundation system. This can be done by ensuring that the downspout routes water away from the wall and the ground slopes away from it.

PILE FOUNDATION

Pile foundations are used to minimize and reduce settlement. There are two classifications of piles: point bearing and friction (Fig. 11). Point bearing piles transmit loads through weak soil to

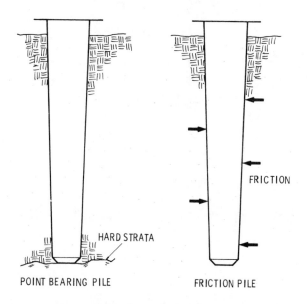

POINT BEARING PILE   FRICTION PILE

Fig. 11. Pile construction.

25

an area that has a better bearing surface. The friction pile depends on the friction between the soil and the pile to support the imposed load.

Many different kinds of materials are used for piles, but the most common are concrete, timber, and steel.

ALL-WEATHER WOOD FOUNDATION

Although it has not yet taken America by storm, the wood foundation may someday do just that. It is composed of wood

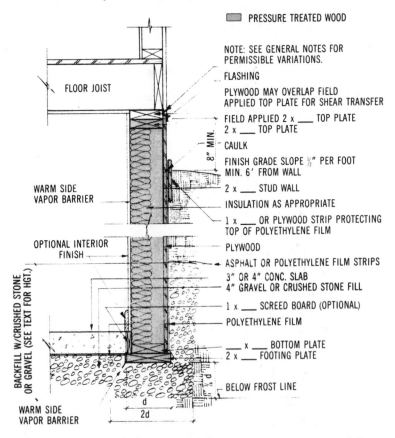

Fig. 12. All-weather wood foundation. Courtesy of National Forest Products Assn.

and plywood soaked with preservatives, and was primarily developed so that a foundation could be installed in cold weather, when concrete cannot.

The wood foundation is not difficult to install, and it is faster to install than a masonry foundation. How well it compares to a masonry foundation over the long run will soon be determined because tests are scheduled to start soon on how well it has fared over the last fifteen years. Fig. 12 gives a detailed look at this foundation.

SUMMARY

There are three general forms of foundations: spread foundations, pile foundations and wood foundations. Spread foundations are the most popular type used. There are three basic types of these: the slab-on-grade, the crawl space, and the basement.

The procedure for constructing a slab-on-grade would be: clear the site, lay out the foundation, place and brace the form boards, add fill, dig the footings, place the base course, place the vapor barrier, reinforce the slab, and reinforce the footings.

The foundation wall, in basement construction, should be built with the utmost care and craftsmanship to resist the assault of ground water.

REVIEW QUESTIONS

1. What are the two general forms of foundations?
2. What is a foundation?
3. Why is base course used?
4. How is a concrete slab cured?
5. How is a basement wall damp-proofed?

CHAPTER 3

Concrete Forms

Since a concrete mixture is semi-fluid, it will take the shape of anything into which it is poured. Molds or forms are necessary to hold the concrete to the required shape until it hardens. Such forms or molds may be made of metal, lumber, or plywood.

The length of time necessary to leave the forms in place depends on the nature of the structure. For small construction work where the concrete bears external weight, the forms may be removed as soon as the concrete will bear its own weight, which is between 12 and 48 hours after the material has been poured. In some instances the concrete is left in place much longer than this.

Forms must be reasonably tight, rigid, and strong enough to sustain the weight of the concrete. They should also be simple and economical and designed so that they may be easily removed and re-erected without damage to themselves or to the concrete. The different shapes into which the concrete is formed mean that each job will present some new problems to be solved.

LUMBER FOR FORMS

Most concrete building construction is done by using wooden or plywood forms. If kiln-dried lumber is used, it should be thoroughly wet before concrete is placed. This is important because the lumber will absorb water from the concrete, and if the forms are made tight (as they should be), the swelling from absorption can cause the forms to buckle or warp. Oiling or using special compounds on the inside of forms (as detailed later) before use is recommended, especially where forms are to be used repeatedly. It prevents absorption of water, assists in keeping them in shape when not in use, and makes their removal from around the concrete much easier.

Spruce or pine seems to be the best all-around material. One or both woods can be obtained in most locations. Hemlock is not usually desirable for concrete form work because it is liable to warp when exposed to concrete. One-inch lumber is normally sufficient for a building form, and so is ⅝-inch, ½-inch, and ¾-inch plywood.

It should be noted that the actual building of forms can be a complicated procedure, particularly when the form is going to be used in building a large structure, such as tall walls or large floors. Such things as the vertical pressure on the form when it is filled with concrete, the lateral pressure, and the amount of deflection of the form must all be considered. In essence, the form is a structure—and one that takes a lot of stress—and therefore it must be built to withstand that stress (Fig. 1).

This granted, there are methods and materials for building forms that carpenters and builders must know.

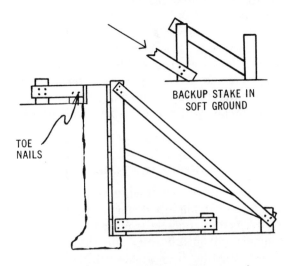

BACKUP STAKE IN
SOFT GROUND

TOE
NAILS

Fig. 1. Cellar wall form in firm ground.

PRACTICALITY

If the job to be done is one on which speed is a prime factor—perhaps because of labor costs—then it is important that the forms not be unduly complicated and time-consuming. The design may be perfectly all right from an engineering standpoint but poor in terms of practicality.

One overall rule is that the components of the form be as large as can be handled practically. When possible, use panels. Panels may be of suitably strong plywood or composed of boards that should all be the same width and fastened together at the back with cleats.

In making forms for columns, the forms may be built as open-ended boxes. Of course, the panels cannot be completely nailed together until they are in place surrounding the column they are reinforcing. It is usually considered good practice to leave a clean-out hole on one side of the bottom of column forms (Fig. 2). Dirt, chips, and other debris are washed out of this hole and then it is closed.

Even when column forms are prepared carefully, though, minor inaccuracies can occur. The form panels may swell, or may move slightly when the concrete is poured. In a well-designed and well-built form such inaccuracies do not matter much because of the allowances built into the forms. As mentioned earlier, forms are wetted down thoroughly, thereby swelling them before use. It is doubly important to do this because of the water-cement ratio. If forms are dry, they will suck water from the concrete and change its properties. A form with a wet surface also solves the problem of wet concrete possibly honeycombing it.

When moldings are desired, they should be recessed into the concrete (Fig. 3). It is very simple to nail triangular strips or other molded shapes on the inside of a form. Projected moldings on concrete require a recess in the form, which is more difficult to make.

Horizontal recessed grooves should be beveled outward at the top and bottom, or at least at the bottom. A form for a flat ledge cannot readily be filled out with poured concrete. Additionally, ledges collect dirt and are likely to become unsightly (Fig. 4).

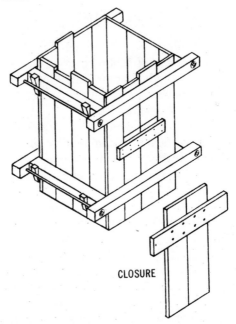

CLOSURE

Fig. 2. Column form with clean-out.

If the placing of concrete can be timed to stop at a horizontal groove, that groove will serve to conceal the horizontal construction joint. One point, often overlooked, is the use of a horizontal construction joint as a ledge to support the forms for the next pour. This is a simple way to hold forms in alignment.

TEXTURE

With forms, a variety of ways may be used to enhance the texture of the finished concrete. If a smooth finish is wanted, forms may be lined with hardboard, linoleum or similar materials. Small-headed nails should be used, and hammer marks should be avoided.

When a rough texture is wanted, rough form boards may be arranged vertically and horizontally, in alternate panels. Other arrangements, such as diamonds, herringbones, or chevrons, can also be used—as can almost anything else the ingenious designer can come up with.

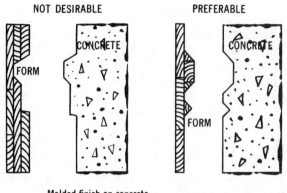

NOT DESIRABLE   PREFERABLE

Molded finish on concrete.

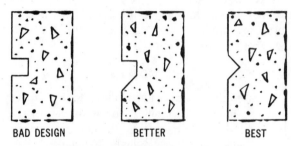

BAD DESIGN   BETTER   BEST

Fig. 3. Horizontal grooves in concrete.

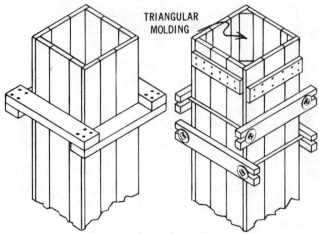

TRIANGULAR MOLDING

Fig. 4. Column form clamps.

33

BRACING

The bracing of concrete formwork falls into a number of categories. The braces that hold wall and column forms in position are usually 1-inch-thick boards or strips. Ordinarily such braces are not heavily stressed because the lateral pressure of the concrete is contained by wall ties and column clamps.

When it is not practical to use wall ties, the braces may be stressed depending on the height of the wall being poured. Braces of this kind are proportioned to support the wall forms against lateral forces. Deep beam and girder forms often require external braces to prevent the side forms from spreading.

The lateral bracing that supports slab forms is also important, not only in terms of safety, but also to prevent the distortions that can occur when the shores are knocked out of position. Lateral bracing should be left in place until the concrete is strong enough to support itself.

ECONOMY

One concern of the builder, particularly today when costs of materials are so high, is economy. Forms for a concrete building, for example, can cost more than the concrete or reinforcing steel or, in some cases, more than both together. Hence it is important to seek out every possible cost-cutting move.

Saving on costs starts with the designer of the building. He or she must keep in mind the forms required for the building's construction and look to build in every possible economy. For example, it may be possible to adjust the size of beams and columns so that they can be formed with a combination of standard lumber or plywood-panel sizes.

Specific example: In making a beam 10 or 12 inches wide, specially ripped form boards are needed. On the other hand, surfaced 1″ × 6″ (5½ inches actual size) will be just right for an 11-inch-wide beam. Experience has shown that it is entirely practical to design structures around relatively standardized beam and column sizes for the sake of economy. When such procedures are followed, any diminution of strength can be made up by increasing the amount of reinforcing steel used — and you will still save money.

Another costly area to avoid is excessive design. It may just require a relatively small amount of concrete to add a certain look to a structure, but the addition can be very expensive in terms of overall cost.

Another economy may be found in the area of lumber lengths. Long lengths can often be used without trimming. Studs need not be cut off at the top of a wall form but can be used in random lengths to avoid waste. Random-length wales can also be used, and where a wall form is built in place, it does no harm if some boards extend beyond the length of the form.

Paradoxically, some very good finish carpenters are not good at form building because they spend a lot of time building the forms—neatness and exact lengths or widths are not required. Because they have overdone the form building, when it comes time to strip the form, there may be many nails to remove and the job may be much more complicated.

FASTENING AND HARDWARE

All sorts of devices to simplify the building and stripping of forms are available. The simplest is the double-headed nail (Fig. 5). The chief advantage of these nails is that they can be pulled out easily because they are only driven into the first head. There are also many different varieties of column clamps, adjustable shores, and screw jacks. Instead of wedges, the screw jacks are especially suitable when solid shores are used.

A number of wedges are usually required in form building. (Fig. 6). Wedges are used to hold form panels in place, draw parts of forms into line, and adjust shores and braces. Usually the carpenter makes the wedges on the job. A simple jig can be rigged up on a table or radial arm saw for the cutting of wedges.

Form ties, available in many styles, are devices that support both sides of wall forms against the lateral pressure of concrete. Used properly, form ties practically eliminate external bracing and greatly simplify the erection of wall forms.

A simple tie is merely a wire that extends through the form, the ends of the wire doubled around a stud or wale on each side. Although low in cost, simple wire ties are not entirely satisfactory because, under pressure from the concrete, they cut into

35

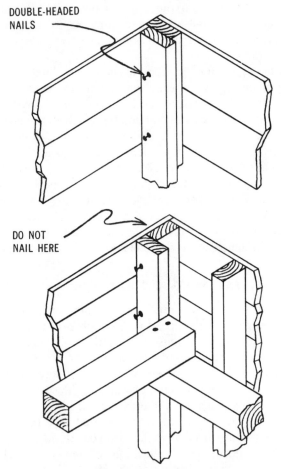

Fig. 5. Double-headed nails in action.
Internal corners make stripping easier.

the wood members and cause irregularities in the wall. The most satisfactory ties can be partially or completely removed from the concrete after it has set and hardened completely.

SIZE AND SPACING

The size and spacing of ties are governed by the pressure of the concrete transmitted through the studs and wales to the ties.

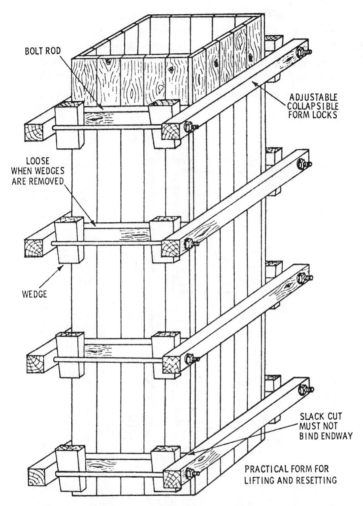

BOLT ROD

ADJUSTABLE
COLLAPSIBLE
FORM LOCKS

LOOSE
WHEN WEDGES
ARE REMOVED

WEDGE

SLACK CUT
MUST NOT
BIND ENDWAY

PRACTICAL FORM FOR
LIFTING AND RESETTING

Fig. 6. Wedges are very important in form building.

In other words, a wale acts as a beam and the ties as reactions. It might be assumed, therefore, that large wales and correspondingly large ties should be used, but in actuality size and spacing are limited by economics. The tie spacing limit is generally considered to be 36 inches with 27 to 30 inches preferred.

Many tie styles make it necessary to place spreaders in the

forms to keep the two sides of the form from being drawn together. Generally, spreaders should be removed so that they are not buried in the concrete. Traditional spreaders are quite good and are readily removed. Fig. 7 shows how wood spreaders are placed and how they are removed.

In more general use are several styles of spreaders combined

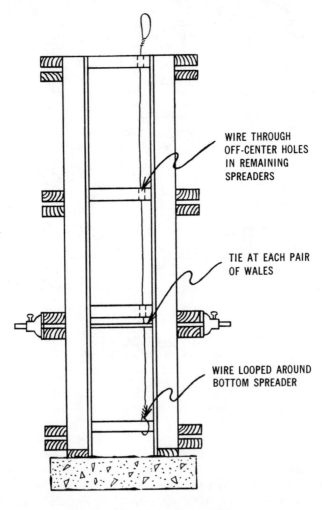

WIRE THROUGH
OFF-CENTER HOLES
IN REMAINING
SPREADERS

TIE AT EACH PAIR
OF WALES

WIRE LOOPED AROUND
BOTTOM SPREADER

Fig. 7. Wall form with spreaders.

with ties. Some of these are made of wire nicked or weakened in such a way that the tie may be broken off in the concrete within an inch or two from the face of the form. After you withdraw the ties, the small holes that remain are easily patched with mortar.

STRIPPING FORMS

Unlike most structures, concrete forms are temporary; the forms must later be removed or stripped. Sometimes, for example in building a single home, forms are used only once and then discarded. But in most cases economy dictates that a form be used and reused. Indeed, economical heavy construction depends on reusing forms.

Because forms are removable, the form designer has certain restrictions. He must not only consider erection but also stripping—the disassembly of the form. Thus, if a form is designed in such a way that final assembly nails are covered, it may be impossible to remove the form without tearing it apart and possibly damaging the partially cured concrete.

Another bad design may be one in which the form is built so that some of its members are encased in concrete and may be difficult to remove. This may result in defective work. It is often advisable to plan column forms so that they can be stripped without disturbing the forms for the beams and girders.

For easier stripping, forms can be coated with special oils. These are not always effective, however, and a number of coating compounds that work well have been developed over the years. These compounds reduce the damage to concrete when stripping is difficult or perhaps carelessly done. The use of these coatings reduces the importance of wetting formwork before placing concrete.

After stripping, forms should be carefully cleaned of all concrete before they are altered and oiled for reuse.

Stripping Forms for Arches

Forms for arches, culverts, and tunnels generally include hinges or loose pieces that, when removed, release the form (Fig. 8). Shores are often set on screw jacks or wedges to simplify their removal. Screw jacks are preferable to wedges because forms held in jacks can be stripped with the least amount of hammering.

For small jobs, where jacks are not available, shores that are almost self-releasing can be made. Two 2 × 6s are fastened into a T-shaped section (Fig. 9) with double-headed nails. When wedged into position, this assembly is a stiff column. After the concrete hardens, the nails are drawn and the column becomes two 2 × 6s, which are relatively easy to remove.

SPECIAL FORMS

Forms are required for building concrete piers, pedestals, and foundations for industrial machinery. The job basically still involves carpentry because such forms differ only in detail from building forms and usually do not have to withstand the pressures that are built up in deep wall or column forms.

Many piers are tapered upward from the footing. In these cases it is necessary to provide a resistance to uplift because the

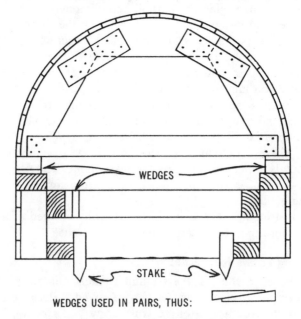

Fig. 8. Soffit form for a small arch culvert.

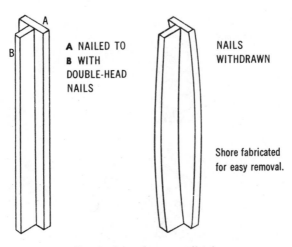

A NAILED TO
B WITH
DOUBLE-HEAD
NAILS

NAILS
WITHDRAWN

Shore fabricated
for easy removal.

Fig. 9. Form for a small job.

semi-liquid concrete tends to float such forms. Once the problem is recognized, it is easily solved. Two or more horizontal planks nailed or wired to spikes will hold down most forms. Sandbags placed on ledgers (Fig. 10) are usually enough for smaller forms.

Some industrial forms are complicated and require cast-in-place hold-down bolts for the machinery. If these bolts are not needed, however, the work is greatly simplified. As Fig. 11 shows, foundations can be built with recesses into which bolts threaded on both ends can be dropped through pipe-lined holes. The machine can be slid onto such a foundation and jacked into position without too much difficulty. If desired, the bolts can be grouted after they are in place.

PREFABRICATED FORMS

In addition to lumber and plywood forms, there is a tremendous variety of prefabricated forms. For example, in Fig. 12 prefab panels are shown. These panels can be made up in different widths and lengths. There is a peg and wedge system to hold the panels together (Fig. 13).

A variety of hinged forms are available, as well as circular ones made of metal (Fig. 14), and clamps (Fig. 15).

41

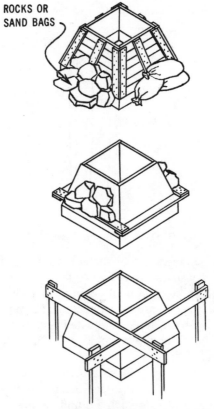

ROCKS OR
SAND BAGS

Fig. 10. Ways of anchoring pedestal footings.

SUMMARY

Concrete mixture is a semi-fluid that will take the shape of any form into which it is poured. Forms are usually made of metal, sheet steel, or wood. Forms must be reasonably tight, rigid, and strong enough to sustain the weight of the concrete.

Most concrete construction, such as building walls, is done by using wooden or plywood forms. Oiling or greasing the inside of the form before use is recommended. This prevents absorption of water from the concrete, which could buckle or warp the forms and weaken the concrete.

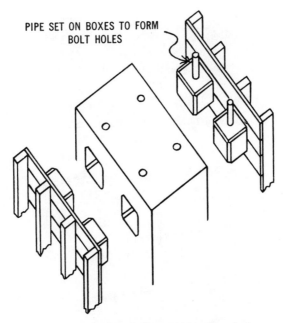

PIPE SET ON BOXES TO FORM
BOLT HOLES

Fig. 11. Concrete form for machine foundation.

Forms may also be prefabricated. Whatever, forms should be regarded as a structure and built with economy and efficiency in mind.

REVIEW QUESTIONS

1. Many times concrete forms are oiled or greased. Why is this done?
2. What are prefab forms?
3. Name two ways to economize when building forms.
4. How much does concrete weigh per cubic foot?

Fig. 12. Prefabricated build-up panels used for concrete forms.

Fig. 13. Peg and wedge system used to connect and hold prefabricated panels together.

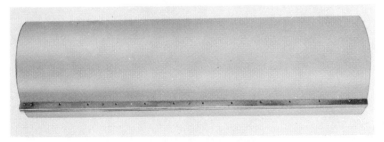

Fig. 14. Circular metal concrete form.

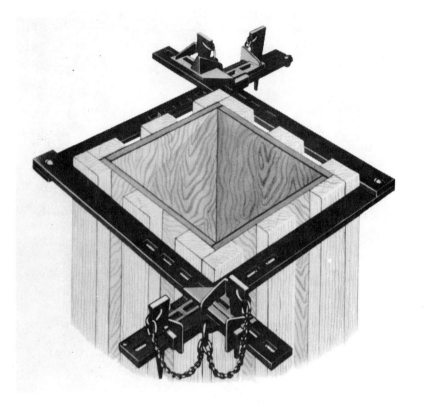

Fig. 15. Spring steel clamps.

CHAPTER 4

Concrete Block Construction

Concrete blocks—or cement blocks, as they are also known —provide suitable building units. By the use of standard-size hollow blocks, basement walls can be erected at a very reasonable cost and are durable, light in weight, fire-resistant and able to carry heavy loads.

The term *concrete masonry* is applied to building units molded from concrete and laid by masons in a wall. The units are made of Portland cement, water, and suitable aggregates such as sand, gravel, crushed stone, cinders, burned shale, or processed slag. The various steps in building concrete masonry walls are approximately as follows:

1. Mixing the mortar.
2. Building the wall between corners.
3. Applying mortar to the blocks.
4. Placing and setting the blocks.
5. Tooling the mortar joints.
6. Building around door and window frames.
7. Placing sills and lintels.
8. Building interior walls.
9. Attaching sills and plates.

WALL THICKNESS

Thickness of concrete masonry walls is usually governed by building codes, if any are in existence at the particular location.

CONCRETE BLOCK CONSTRUCTION

Eight inches is generally specified as the minimum thickness for all exterior walls, and for load-bearing interior walls. Partitions and curtain walls are often made 3, 4, or 6 inches thick. The thickness of bearing walls in heavily loaded buildings is properly governed by the loading. Allowable working loads are commonly 70 pounds per square inch (70 p.s.i.) of gross wall area when laid in 1:1:6 (1 volume of Portland cement, 1 volume of lime putty or hydrated lime, and 6 parts damp, loose mortar sand). On small jobs, masonry cement can be purchased and the mortar mixed with 1 part masonry cement to 3 parts sand (mix dry before adding water). A mortar mix can also be obtained to which all you add is water. When a mortar of maximum strength is desired for use in load-bearing walls, or walls subjected to heavy pressure and subjected to freezing and thawing, a mortar made of 1 part Portland cement, 1 part masonry cement, and not more than 6 parts sand is recommended.

Shapes and Sizes of Standard Masonry Units

Concrete masonry units, often referred to as concrete blocks, are made in several sizes and shapes. The shapes and sizes of the most generally used units are shown in Figs. 1 and 2. The 7¾

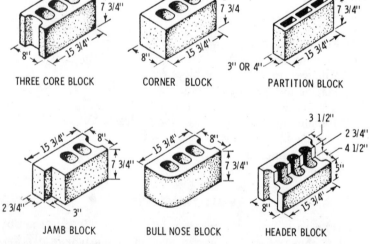

Fig. 1. Common sizes and shapes of concrete masonry blocks.

Fig. 2. Two core masonry block being installed. Courtesy of Bilco.

× 8 × 15¾-inch unit (usually considered 8 × 8 × 16) is the size most commonly used. The 8 × 8 × 16-inch size laid up in a single thickness produces a wall 8 inches thick with courses 8 inches in height. One course is equivalent to three brick courses.

Actual heights and lengths of the block are usually about ¼ to ⅜ inch less to allow for the thickness of the mortar joints, as illustrated in Fig. 3. A special corner block has one finished square end for use at the corners of the wall. A special jamb block for window and door frames is also made, as illustrated. All units shown are made in half lengths and quarters, as well as in full lengths, to permit breaking the vertical joints in alternate courses. Units are also made in widths of approximately 4, 6, and 12 inches.

Concrete made with sand, gravel, and crushed stone weighs

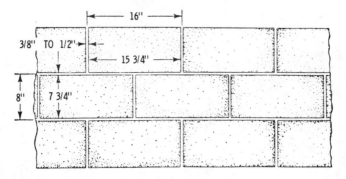

Fig. 3. Typical dimensions of mortar joints when using concrete blocks as building units.

approximately 145 pounds per cubic foot. Concrete blocks made from such aggregates are known as heavyweight units. Those made with cinders, burned shale, and processed slags weigh approximately two-thirds as much as heavyweight units and have a greater heat-insulating value. It may be advisable to use lightweight units where heat comfort is a major consideration, but they do not have the strength of heavyweight units. Fig. 4 shows a number of designs that can be obtained by using different size concrete blocks.

WALL CONSTRUCTION

Concrete masonry walls that are subject to average loading and exposure should be laid up with mortar composed of 1 volume of masonry cement and between 2 and 3 volumes of damp, loose mortar sand; or 1 volume of Portland cement, between 1 and 1¼ volumes of hydrated lime or lime putty, and between 4 and 6 volumes of damp, loose mortar sand. Enough water is added to produce a workable mixture.

Concrete masonry walls and isolated piers that are subject to severe conditions, such as extremely heavy loads, violent winds, earthquakes, severe frost action, or other conditions requiring extra strength, should be laid up with mortar composed of 1 volume of masonry cement, plus 1 volume of Portland cement, and between 4 and 6 volumes of damp, loose mortar sand; or 1 volume of Portland cement, between 2 and 3 volumes of damp,

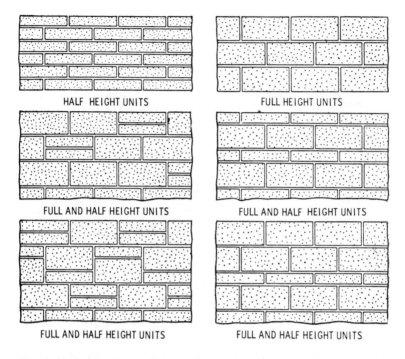

HALF HEIGHT UNITS

FULL HEIGHT UNITS

FULL AND HALF HEIGHT UNITS

FULL AND HALF HEIGHT UNITS

FULL AND HALF HEIGHT UNITS

FULL AND HALF HEIGHT UNITS

Fig. 4. Various patterns that can be produced with standard concrete block units.

loose mortar sand and up to ¼ volume of hydrated lime or lime putty. Enough water is added to produce a workable mixture.

Laying Blocks at Corners

In laying up corners with concrete masonry blocks, place a taut line all the way around the foundation, with the ends of the string tied together. It is customary to lay up the corner blocks three or four courses high and use them as guides in laying the rest of the walls.

A full width of mortar is placed on the footing, as shown in Fig. 5, with the first course laid 2 or 3 blocks long each way from the corner. The second course is a half block shorter each way than the first course, the third course is a half block shorter than the second, etc. Thus the corners are stepped back until only the corner blocks are laid. Use a line and level frequently to see that

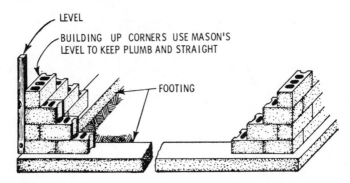

Fig. 5. Laying up corners when building with concrete masonry block units.

the blocks are laid straight and that the corners are plumb. It is customary that such special units as corner blocks, door- and window-jamb blocks, fillers, and veneer blocks be provided prior to starting the laying of the blocks.

Building Walls Between Corners

In laying walls between corners, a line is stretched tightly from corner to corner to serve as a guide (Fig. 6). The line is fastened to nails or wedges driven into the mortar joints so that, when stretched, it just touches the upper outer edges of the block laid in the corners. The blocks in the wall between corners are laid so that they will just touch the cord in the same manner. In this way, straight horizontal joints are secured. Prior to laying up the outside wall, the door and window frames should be on hand to set in place as guides for obtaining the correct opening.

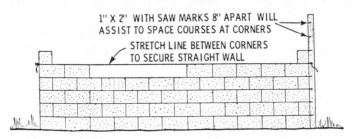

Fig. 6. Showing procedure in laying concrete block walls.

Applying Mortar to Blocks

The usual practice is to place the mortar in two separate strips, both for the horizontal or bed joints and for the vertical or end joints, as shown in Fig. 7. The mortar is applied only on the face shells of the block. This is known as face-shell bedding. The air spaces thus formed between the inner and outer strips of mortar will help produce a dry wall.

Masons often stand the block on end and apply mortar for the end joint. Sufficient mortar is put on to make sure that the joint will be well filled. Some masons apply mortar on the end of the block previously laid, as well as on the end of the block to be laid next to it, to make certain that the vertical joint will be completely filled—have no air spaces.

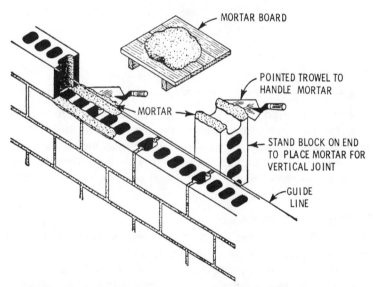

Fig. 7. The usual practice in applying mortar to concrete blocks.

Placing and Setting Blocks

In placing, the block that has mortar applied to one end is then picked up, as shown in Fig. 8, and shoved firmly against the block previously placed. Note that the mortar is already in place in the bed or horizontal joints.

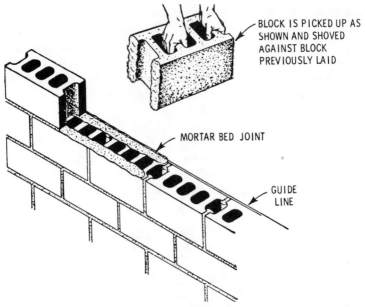

BLOCK IS PICKED UP AS SHOWN AND SHOVED AGAINST BLOCK PREVIOUSLY LAID

MORTAR BED JOINT

GUIDE LINE

Fig. 8. The common method of picking up and setting concrete block.

Mortar squeezed out of the joints is carefully cut off with the trowel and applied on the other end of the block, or thrown back onto the mortar board for use later. The blocks are laid to touch the guide line, and tapped with the trowel to get them straight and level, as shown in Fig. 9. In a well constructed wall, mortar joints will average ⅜-inch thick.

Tooling Mortar Joints

When the mortar has become firm, the joints are tooled, or finished. Various tools are used for this purpose. A rounded or V-shaped steel jointer is the tool most commonly used, as shown in Fig. 10. Tooling compresses the mortar in the joints, forcing it up tightly against the edges of the block, and leaves the joints smooth and watertight. Some joints may be cut off flush and struck with the trowel.

Building Around Door and Window Frames

There are several acceptable methods of building door and window frames in concrete masonry walls. One method used is

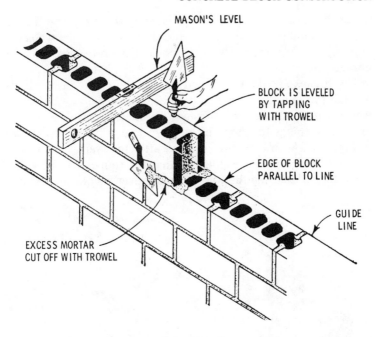

MASON'S LEVEL

BLOCK IS LEVELED
BY TAPPING
WITH TROWEL

EDGE OF BLOCK
PARALLEL TO LINE

GUIDE
LINE

EXCESS MORTAR
CUT OFF WITH TROWEL

Fig. 9. A method of laying concrete blocks. Good workmanship requires straight courses with face of wall plumb.

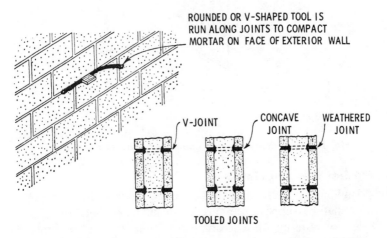

ROUNDED OR V-SHAPED TOOL IS
RUN ALONG JOINTS TO COMPACT
MORTAR ON FACE OF EXTERIOR WALL

V-JOINT   CONCAVE   WEATHERED
           JOINT   JOINT

TOOLED JOINTS

Fig. 10. A method of tooling to compact mortar between concrete blocks.

to set the frames in the proper position in the wall. The frames are then plumbed and carefully braced, after which the walls are built up against them on both sides. Concrete sills may be poured later.

The frames are often fastened to the walls with anchor bolts passing through the frames and embedded in the mortar joints. Another method of building frames in concrete masonry walls is to build openings for them, using special jamb blocks, as shown in Fig. 11. The frames are inserted after the wall is built. The only advantage to this method is that the frames can be taken out without damaging the wall, should it ever become necessary.

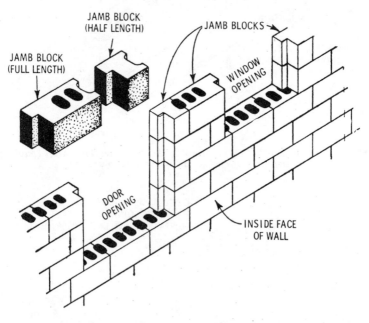

Fig. 11. A method of laying openings for doors and windows.

Placing Sills and Lintels

Building codes require that concrete block walls above openings shall be supported by arches or lintels of metal or masonry (plain or reinforced). The arches or lintels must extend into the walls not less than 4 inches on each side. Stone or other nonrein-

56

forced masonry lintels should not be used unless supplemented on the inside of the wall with iron or steel lintels. Fig. 12 illustrates typical methods of inserting concrete reinforced lintels to provide for door and window openings.

These are usually prefabricated, but may be made up on the job if desired. Lintels are reinforced with steel bars placed 1½ inches from the lower side. The number and size of reinforcing rods depend upon the width of the opening and the weight of the load to be carried.

Sills serve the purpose of providing watertight bases at the bottom of wall openings. Since they are made in one piece, there are no joints for possible leakage of water into the walls below. They are sloped on the top face to drain water away quickly. They are usually made to project 1½ to 2 inches beyond the wall face, and are made with a groove along the lower outer edge to provide a drain so that water dripping off the sill will fall free and not flow over the face of the wall causing possible staining.

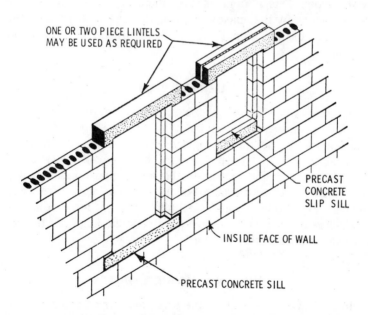

ONE OR TWO PIECE LINTELS MAY BE USED AS REQUIRED

PRECAST CONCRETE SLIP SILL

INSIDE FACE OF WALL

PRECAST CONCRETE SILL

Fig. 12. A method of inserting precast concrete lintels and sills in concrete block walls.

Slip sills are popular because they can be inserted after the wall has been built and therefore require no protection during construction. Since there is an exposed joint at each end of the sill, special care should be taken to see that it is completely filled with mortar and the joints packed tight.

Lug sills are projected into the concrete block wall (usually 4 inches at each end). The projecting parts are called lugs. There are no vertical mortar joints at the juncture of the sills and the jambs. Like the slip sill, lug sills are usually made to project from 1½ to 2 inches over the face of the wall. The sill is provided with a groove under the lower outer edge to form a drain. Frequently they are made with washes at either end to divert water away from the juncture of the sills and the jambs. This is in addition to the outward slope on the sills.

At the time the lug sills are set, only the portion projecting into the wall is bedded in mortar. The portion immediately below the wall opening is left open and free of contact with the wall below. This is done in case there is minor settlement or adjustments in the masonry work during construction, thus avoiding possible damage to the sill during the construction period.

BASEMENT WALLS

These walls should not be narrower than the walls immediately above them, and not less than 12 inches for unit masonry walls. Solid cast-in-place concrete walls are reinforced with at least one ⅜-inch deformed bar (spaced every 2 feet) continuous from the footing to the top of the foundation wall. Basement walls with 8-inch hollow concrete blocks frequently prove very troublesome. All hollow-block foundation walls should be capped with a 4-inch solid concrete block, or the core should be filled with concrete.

BUILDING INTERIOR WALLS

Interior walls are built in the same manner as exterior walls. Load-bearing interior walls are usually made 8 inches thick; partition walls that are not load bearing are usually 4 inches thick. The recommended method of joining interior load-bearing walls to exterior walls is illustrated in Fig. 13.

BUILDING TECHNIQUES

Sills and plates are usually attached to concrete block walls by means of anchor bolts, as shown in Fig. 14. These bolts are placed in the cores of the blocks, and the cores are filled with concrete. The bolts are spaced about 4 feet apart under average conditions. Usually ½-inch bolts are used and should be long enough to go through two courses of blocks and project through the plate about an inch to permit the use of a large washer and the anchor-bolt nut.

Installation of Heating and Ventilating Ducts

These ducts are provided for as shown on the architect's plans. The placement of the heating ducts depends upon the type of wall, whether it is load bearing or not. A typical example of

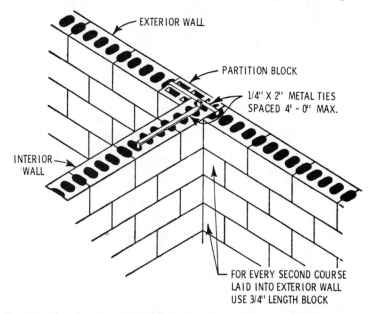

EXTERIOR WALL

PARTITION BLOCK

1/4" X 2" METAL TIES SPACED 4' - 0" MAX.

INTERIOR WALL

FOR EVERY SECOND COURSE LAID INTO EXTERIOR WALL USE 3/4" LENGTH BLOCK

Fig. 13. Showing detail of joining interior and exterior wall in concrete block construction.

CONCRETE BLOCK CONSTRUCTION

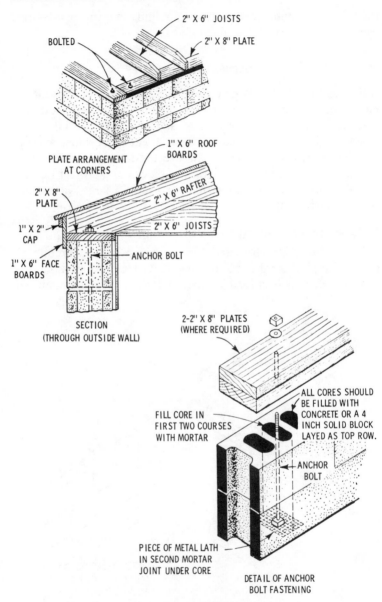

2" X 6" JOISTS

2" X 8" PLATE

BOLTED

PLATE ARRANGEMENT
AT CORNERS

1" X 6" ROOF
BOARDS

2" X 8"
PLATE

2" X 6" RAFTER

1" X 2"
CAP

2" X 6" JOISTS

1" X 6" FACE
BOARDS

ANCHOR BOLT

SECTION
(THROUGH OUTSIDE WALL)

2-2" X 8" PLATES
(WHERE REQUIRED)

ALL CORES SHOULD
BE FILLED WITH
CONCRETE OR A 4
INCH SOLID BLOCK
LAYED AS TOP ROW.

FILL CORE IN
FIRST TWO COURSES
WITH MORTAR

ANCHOR
BOLT

PIECE OF METAL LATH
IN SECOND MORTAR
JOINT UNDER CORE

DETAIL OF ANCHOR
BOLT FASTENING

*Fig. 14. Showing details of methods used to anchor sills and plates to
concrete block walls.*

placing the heating or ventilating ducts in an interior concrete masonry wall is shown in Fig. 15.

Interior concrete block walls that are not load bearing, and that are to be plastered on both sides, are frequently cut through to provide for the heating duct, the wall being flush with the ducts on either side. Metal lath is used over the ducts.

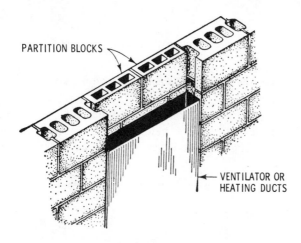

PARTITION BLOCKS

VENTILATOR OR HEATING DUCTS

Fig. 15. Method of installing ventilating and heating duct in concrete block walls.

Electrical Outlets

These outlets are provided for by inserting outlet boxes in the walls as shown in Fig. 16. All wiring should be installed to conform with the requirements of the National Electrical Code or local codes.

Insulation

Block walls can (and should) be insulated in a number of ways. One way is to secure studs to the walls with cut nails or other fasteners, then staple batt insulation to the studs. But a simpler way is with Styrofoam insulation. This comes in easy-to-cut sheets. It is applied to the interior walls with mastic and then, because it is flammable, covered with gypsum drywall or other noncombustible material. Fig. 17 shows this.

The Styrofoam may also be used on the exterior of the wall.

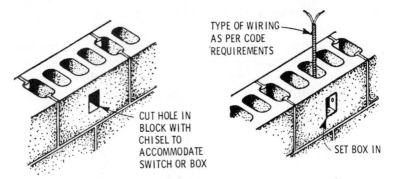

TYPE OF WIRING
AS PER CODE
REQUIREMENTS

CUT HOLE IN
BLOCK WITH
CHISEL TO
ACCOMMODATE
SWITCH OR BOX

SET BOX IN

Fig. 16. Method showing installation of electrical switches and outlet boxes in concrete block walls.

In this case it is secured only to that part of the foundation above the earth and one foot below. The Styrofoam, in turn, is covered with a cement or other mixture as specified by the manufacturer.

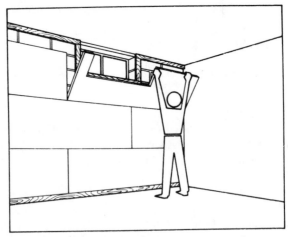

Fig. 17. Styrofoam rigid insulation board can be glued to masonry.

Flashing

Adequate flashing with rust and corrosion-resisting material is of the utmost importance in masonry construction, since it prevents water from getting through the walls at vulnerable points.

Points requiring protection by flashing are: tops and sides of projecting trim; under coping and sills; at intersection of wall and roof; under built-in gutters; at intersection of chimney and roof; and at all other points where moisture is likely to gain entrance. Flashing material usually consists of No. 26 gauge (14-oz.) copper sheet or other approved noncorrodible material.

BLOCK AND BRICK WALLS

Concrete blocks and face bricks provide a suitable combination. Cost can often be reduced by laying the blocks with only the front tier of brick, and using clay tile or concrete blocks for the inner tiers. Fig. 18 illustrates the method of laying hollow concrete blocks in backing up an 8- and 12-inch wall. Concrete blocks laid in this manner provide a solid and economical wall.

Since common face brick units are approximately 2⅔″ × 3¾″ × 8″ in size, it requires three brick courses to one course of the concrete blocks. This allows for cross bonding every seventh course. Most building codes require the facing and backing of concrete block walls to be bonded with either at least one full header course in each seven courses, or with at least one full length header in each 1½ square feet of wall surface. The distance between adjacent full-length headers should not, in any event, exceed 20 inches either vertically or horizontally. Where two or more hollow units are used to make up the thickness of a wall, the stretcher courses should be bonded at vertical intervals not exceeding 34 inches, by lapping at least 3¾ inches over the unit below.

In cavity walls, the facing and backing should be securely tied together with suitable bonding ties of adequate strength, as shown in Fig. 19. A steel rod, ³/₁₆-inch in diameter, or a metal tie of equivalent stiffness (coated with a non-corroding metal or other approved protective coating), should be used for each 3 square feet of wall surface. Where hollow masonry units are laid with the cells vertical, rectangular ties shall be used; in other walls, the ends of the ties shall be bent to 90° angles to provide hooks not less than 2 inches long. Ties should be embedded in the horizontal joints of the facing and backing. Additional bonding ties should be provided at all openings, spaced not more than

CONCRETE BLOCK CONSTRUCTION

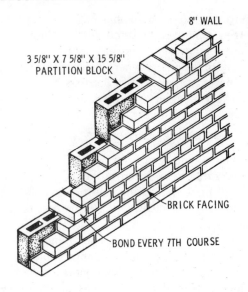

8" WALL

3 5/8" X 7 5/8" X 15 5/8"
PARTITION BLOCK

BRICK FACING

BOND EVERY 7TH COURSE

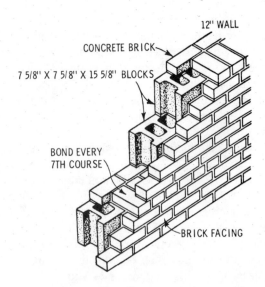

12" WALL

CONCRETE BRICK

7 5/8" X 7 5/8" X 15 5/8" BLOCKS

BOND EVERY
7TH COURSE

BRICK FACING

Fig. 18. A method of building combination masonry walls with brick and 4- and 8-inch concrete blocks.

3 feet apart around the perimeter and within 12 inches of the opening.

FLOORS

In concrete masonry construction, floors may be made entirely of wood or concrete, although a combination of concrete and wood is sometimes used. Wooden floors (Fig. 20) must be framed with one governing consideration: they must be made strong enough to carry the load. The type of building and its use will determine the type of floor used, the thickness of the sheathing, and approximate spacing of the joists. The girder is usually made of heavy timber and is used to support the lighter joists between the outside walls.

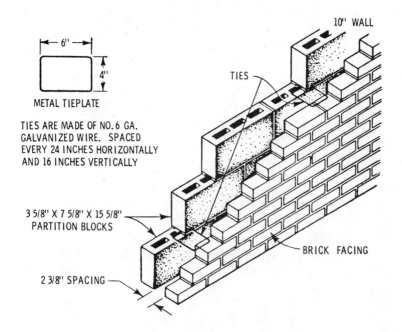

6"

4"

METAL TIEPLATE

TIES ARE MADE OF NO. 6 GA. GALVANIZED WIRE. SPACED EVERY 24 INCHES HORIZONTALLY AND 16 INCHES VERTICALLY

3 5/8" X 7 5/8" X 15 5/8" PARTITION BLOCKS

2 3/8" SPACING

10" WALL

TIES

BRICK FACING

Fig. 19. Metal ties used in bonding masonry walls of concrete block and brick.

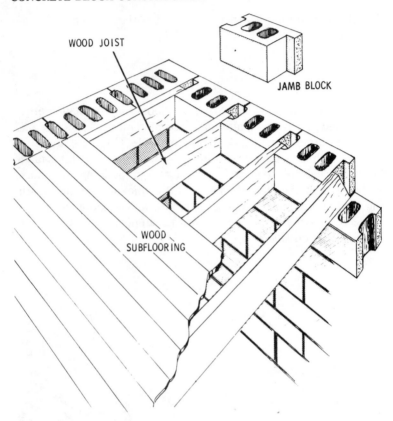

WOOD JOIST

JAMB BLOCK

WOOD SUBFLOORING

Fig. 20. Method of installing wooden floor joists in concrete blocks.

Concrete Floors

There are usually two types of concrete floors—cast-in-place concrete, and precast-joists used with cast-in-place or precast concrete units.

Cast-in-Place—Cast-in-place concrete floors, such as shown in Fig. 21, are used in basements, and in residences without basements, and are usually reinforced by means of wire mesh to provide additional strength and to prevent cracks in the floor. This type of floor is used for small areas. It used to be popular for porch floors.

Another type of cast-in-place concrete floor is shown in Fig.

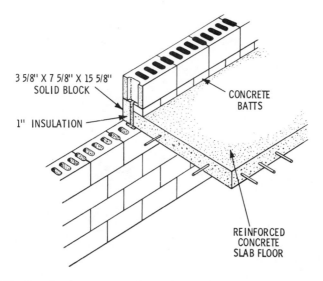

3 5/8" X 7 5/8" X 15 5/8"
SOLID BLOCK

1" INSULATION

CONCRETE
BATTS

REINFORCED
CONCRETE
SLAB FLOOR

Fig. 21. Cast-in-place concrete floor supported by the concrete block wall.

22. This floor has more strength because of the built-in joist, and can be used for larger areas.

Precast Joists — Fig. 23 shows the precast-joist type. The joists are set in place, the wooden forms inserted, the concrete floor is poured, and the forms are removed. Another type of precast-joist concrete floor (Fig. 24) consists of precast concrete joists covered with cast-in-place concrete slabs. The joists are usually made in a concrete products plant.

Usual spacing of the joists is from 27 to 33 inches, depending upon the span and the load. The cast-in-place concrete slab is usually 2 or 2½ inches thick, and extends down over the heads of the joists about ½ inch. Precast concrete joists are usually made in 8-, 10-, and 12-inch depths. The 8-inch joists are used for spans up to 16 feet, the 10-inch joists are for spans between 16 and 20 feet, and the 12 inch joists are for spans from 20 to 24 feet.

Where masonry partitions that are not load bearing are placed parallel to the joists, it is customary to double the joists under the partition. If the partition runs at right angles to the joists, the usual practice is to design the floor to carry an additional load of

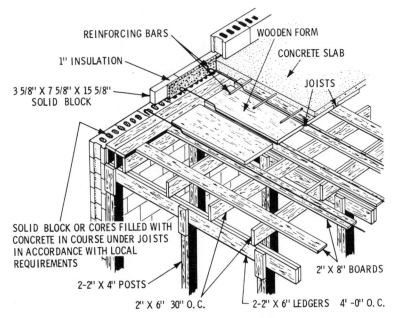

REINFORCING BARS

WOODEN FORM

CONCRETE SLAB

1" INSULATION

3 5/8" X 7 5/8" X 15 5/8"
SOLID BLOCK

JOISTS

SOLID BLOCK OR CORES FILLED WITH
CONCRETE IN COURSE UNDER JOISTS
IN ACCORDANCE WITH LOCAL
REQUIREMENTS

2-2" X 4" POSTS

2" X 6" 30" O. C.

2-2" X 6" LEDGERS   4' -0" O. C.

2" X 8" BOARDS

Fig. 22. Framing construction for cast-in-place concrete floor.

20 pounds per square foot. Precast concrete joists may be left exposed on the underside and painted, or a suspended ceiling may be used. An attractive variation in exposed joist treatment is to double the joists and increase the spacing. Where this is done, the concrete slab is made 2½ inches thick for spacings up to 48 inches, and 3 inches thick for spacings from 48 to 60 inches. Joists may be doubled by setting them close together or by leaving a space between them and filling the space with concrete.

Wood Floors

Where wood-surfaced floors are used in homes, any type of hardwood, such as maple, birch, beech, or oak, may be laid over the structural concrete floor. The standard method of laying hardwood flooring over a structural concrete floor slab is to nail the boards to 2″ × 2″ or 2″ × 3″ sleepers. These sleepers may be tied to the tops of the stirrups protruding from the precast

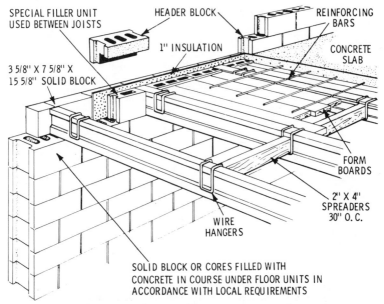

SPECIAL FILLER UNIT
USED BETWEEN JOISTS

HEADER BLOCK

REINFORCING
BARS

1" INSULATION

CONCRETE
SLAB

3 5/8" X 7 5/8" X
15 5/8" SOLID BLOCK

FORM
BOARDS

2' X 4"
SPREADERS
30" O. C.

WIRE
HANGERS

SOLID BLOCK OR CORES FILLED WITH
CONCRETE IN COURSE UNDER FLOOR UNITS IN
ACCORDANCE WITH LOCAL REQUIREMENTS

Fig. 23. Illustrating framing of precast concrete floor joists.

concrete joists. They should be placed not over 16 inches on center, preferably 12 inches in residential construction. Before the hardwood flooring is nailed to the sleepers, the concrete floor must be thoroughly hardened and free from moisture. It is also best to delay laying the hardwood floors until after drywall installers or plasterers have finished work.

Certain types of parquet and design wood floorings are laid directly on the concrete, being bonded to the surface with adhesive. The concrete surface should be troweled smooth and be free from moisture. No special topping is required; the ordinary level sidewalk finish provides a satisfactory base. Manufacturers' directions for laying this type of wood flooring should be followed carefully in order to ensure satisfactory results in the finished floor.

SUMMARY

Concrete blocks provide a very good building material. Blocks are very reasonable in cost, are durable, lightweight, fire resis-

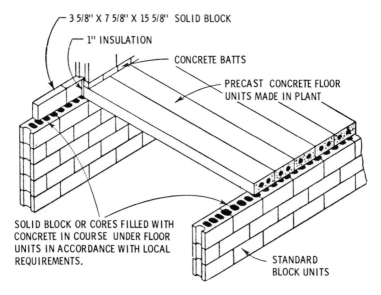

3 5/8" X 7 5/8" X 15 5/8" SOLID BLOCK

1" INSULATION

CONCRETE BATTS

PRECAST CONCRETE FLOOR UNITS MADE IN PLANT

SOLID BLOCK OR CORES FILLED WITH CONCRETE IN COURSE UNDER FLOOR UNITS IN ACCORDANCE WITH LOCAL REQUIREMENTS.

STANDARD BLOCK UNITS

Fig. 24. *Method of framing cored concrete floor units into concrete walls.*

tant and can carry heavy loads. The thickness of concrete blocks is generally 8 inches (except for partition blocks), which is generally specified as the minimum thickness for all exterior walls and for load-bearing interior walls. Partitions and curtain walls are often made 3, 4, or 6 inches thick.

Concrete blocks are made in several sizes and shapes. For various openings, such as windows or doors, a jamb block is generally used to adapt to various frame designs. Building codes require that concrete block walls above openings shall be supported by arches of steel or precast masonry lintels. Lintels are usually prefabricated, but may be made on the job if desired. They are reinforced with steel bars placed approximately 1½ inches from the lower side.

Brick and concrete blocks provide a suitable combination, and the cost of a wall can be reduced by laying the blocks with the bricks into some very attractive patterns. A solid 8- or 12-inch wall can be constructed in this manner at an economical cost.

In concrete block construction, floors may be entirely of wood or concrete, although a combination of the two materials is sometimes used. Joists may be installed in the block wall by using jamb blocks or by installing a sill plate at the top of the wall. Concrete floors are generally of the cast-in-place, precast units, or precast-joist types.

REVIEW QUESTIONS

1. What is the general thickness of concrete blocks?
2. What are the standard measurements of a full-size concrete block?
3. What are lintels? Why are they used?
4. How are combination concrete block and brick walls constructed?
5. What are precast joists?

CHAPTER 5

Framing

KNOWLEDGE OF LUMBER

The basic construction material in carpentry is lumber. There are many kinds of lumber and they vary greatly in structural characteristics. Here, we deal with lumber common to construction carpentry, its application, the standard sizes in which it is available, and the methods of computing lumber quantities in terms of board feet.

Standard Sizes of Bulk Lumber

Lumber is usually sawed into standard lengths, widths, and thicknesses. This permits uniformity in planning structures and in ordering material. Table 1 lists the common widths and thicknesses of wood in rough and in dressed dimensions in the United States. Standards have been established for dimension differences between nominal size and the standard size (which is actually the reduced size when dressed). It is important that these dimension differences be taken into consideration when planning a structure. A good example of the dimension differences may be illustrated by the common 2 × 4. As may be seen in the table, the familiar quoted size (2 × 4) refers to a rough or nominal dimension, but the actual standard size to which the lumber is dressed is 1½″ × 3½″.

Grades of Lumber

Select Lumber—Select lumber is of good appearance and finished or dressed. It is identified by the following grade names:

TABLE 1. Your Guide to New Sizes of Lumber

WHAT YOU ORDER	WHAT YOU GET		WHAT YOU USED TO GET
	*Dry or Seasoned	**Green or Unseasoned	Seasoned or Unseasoned
1 x 4	¾ x 3½	²⁵⁄₃₂ x 3⁹⁄₁₆	²⁵⁄₃₂ x 3⅝
1 x 6	¾ x 5½	²⁵⁄₃₂ x 5⅝	²⁵⁄₃₂ x 5½
1 x 8	¾ x 7¼	²⁵⁄₃₂ x 7½	²⁵⁄₃₂ x 7½
1 x 10	¾ x 9¼	²⁵⁄₃₂ x 9½	²⁵⁄₃₂ x 9½
1 x 12	¾ x 11¼	²⁵⁄₃₂ x 11½	²⁵⁄₃₂ x 11½
2 x 4	1½ x 3½	1⁹⁄₁₆ x 3⁹⁄₁₆	1⅝ x 3⅝
2 x 6	1½ x 5½	1⁹⁄₁₆ x 5⅝	1⅝ x 5½
2 x 8	1½ x 7¼	1⁹⁄₁₆ x 7½	1⅝ x 7½
2 x 10	1½ x 9¼	1⁹⁄₁₆ x 9½	1⅝ x 9½
2 x 12	1½ x 11¼	1⁹⁄₁₆ x 11½	1⅝ x 11½
4 x 4	3½ x 3½	3⁹⁄₁₆ x 3⁹⁄₁₆	3⅝ x 3⅝
4 x 6	3½ x 5½	3⁹⁄₁₆ x 5⅝	3⅝ x 5½
4 x 8	3½ x 7¼	3⁹⁄₁₆ x 7½	3⅝ x 7½
4 x 10	3½ x 9¼	3⁹⁄₁₆ x 9½	3⅝ x 9½
4 x 12	3½ x 11¼	3⁹⁄₁₆ x 11½	3⅝ x 11½

*19% Moisture Content or under.
**Over 19% Moisture Content.

Grade B and better. Grade B is suitable for natural finishes, of high quality, and is generally clear.

Grade C. Grade C is adapted to high-quality paint finish.

Grade D. Grade D is suitable for paint finishes and is between the higher finishing grades and the common grades.

Common Lumber—Common lumber is suitable for general construction and utility purposes, and is identified by the following grade name:

No. 1 common. No. 1 common is suitable for use without waste. It is sound and tight knotted, and may be considered watertight material.

No. 2 common. No. 2 common is less restricted in quality than No. 1, but of the same general quality. It is used for framing, sheathing, and other structural forms where the stress or strain is not excessive.

No. 3 common. No. 3 common permits some waste with prevailing grade characteristics larger than in No. 2. It is used for such rough work as footings, guardrails, and rough subflooring.

No. 4 common. No. 4 common permits waste, and is of low quality, admitting the coarsest features, such as decay and holes. It is used for sheathing, subfloors, and roof boards in the cheaper types of construction. The most important industrial outlet for this grade is for boxes and shipping crates.

No. 5 common. No. 5 common is not produced in some species. The only requirement is that it must be usable for boxes and crates.

FRAMING LUMBER

The frame of a building consists of the wooden form constructed to support the finished members of the structure. It includes such items as posts, girders (beams), joists, subfloor, sole plate, studs, and rafters. Softwoods are usually used for light-wood framing and all other aspects of construction carpentry considered in this book. They are cut into the standard sizes required for light framing, including 2 × 4, 2 × 6, 2 × 8, 2 × 10, 2 × 12, and all other sizes required for framework, with the exception of those sizes classed as structural lumber.

Although No. 1 and No. 3 common are sometimes used for framing, No. 2 common is most often used, and is therefore most often stocked and available in retail lumberyards in the common sizes for various framing members. However, the size of lumber required for any specific structure will vary with the design of the building, such as light-frame or heavy-frame, and the design of the particular members such as beams or girders.

The exterior walls of a frame building usually consist of three layers—sheathing, building paper, and siding. Sheathing lumber is usually 1 × 6s or 1 × 8s, No. 2 or No. 3 common softwood. It may be plain, tongue-and-groove, or ship lapped. The siding lumber may be grade C, which is most often used. Siding is usually procured in bundles consisting of a given number of square feet per bundle, and comes in various lengths up to a maximum of 20 feet. Sheathing grade (CD) plywood is also extensively used.

COMPUTING BOARD FEET

The arithmetic method of computing the number of board feet in one or more pieces of lumber is by the use of the following formula:

$$\frac{\text{Pieces} \times \text{Thickness (inches)} \times \text{Width (inches)} \times \text{Length (feet)}}{12}$$

Example 1—Find the number of board feet in a piece of lumber 2 inches thick, 10 inches wide, and 6 feet long.

$$\frac{1 \times 2 \times 10 \times 6}{12} = 10 \text{ board feet.}$$

Example 2—Find the number of board feet in 10 pieces of lumber 2 inches thick, 10 inches wide, and 6 feet long.

$$\frac{10 \times 2 \times 10 \times 6}{12} = 100 \text{ board feet.}$$

Example 3—Find the number of board feet in a piece of lumber 2 inches thick, 10 inches wide, and 18 inches long.

$$\frac{2 \times 10 \times 18}{144} = 2\frac{1}{2} \text{ board feet.}$$

(NOTE: If all three dimensions are expressed in inches, the same formula applies except the divisor is changed to 144.)

The tabular method of computing board feet with use of a framing square is covered in Chapter 23, "How To Use the Steel Square" in *Carpenters and Builders Library Vol. 1.*

METHODS OF FRAMING

Good material and workmanship will be of very little value unless the underlying framework of a building is strong and rigid. The resistance of a house to such forces as tornadoes and earthquakes, and control of cracks due to settlement, all depends on a good framework.

Although it is true that no two buildings are put together in exactly the same manner, disagreement exists among architects and carpenters as to which method of framing will prove most satisfactory for a given condition. *Light framed construction* may be classified into three distinct types known as:

1. Balloon frame.
2. Plank and beam.
3. Platform frame.

Balloon Frame Construction

The principal characteristic of balloon framing is the use of studs extending in one piece from the foundation to the roof, as shown in Fig. 1. The joists are nailed to the studs and also supported by a ledger board set into the studs. Diagonal sheathing may be used instead of wall board to eliminate corner bracing.

Plank and Beam Construction

The plank and beam construction is said to be the oldest method of framing in this country, having been imported from England in colonial times. It is still used, particularly where the builder wants large open areas in a house. This type of framing is characterized by heavy timber posts, as shown in Fig. 2, and often with intermediate posts between.

Platform Frame Construction

Platform framing is characterized by platforms independently framed, the second or third floor being supported by the studs from the first floor, as shown in Fig. 3. The chief advantage in this type of framing (in all-lumber construction) lies in the fact that if there is any settlement due to shrinkage, it will be uniform throughout and will not be noticeable.

FOUNDATION SILLS

The foundation sill consists of a plank or timber resting upon the foundation wall. It forms the support or bearing surface for the outside of the building, and as a rule, the first floor joists rest upon it. Shown in Fig. 4 is the balloon-type construction of framing for the first floor. In Fig. 5 is shown one way to anchor, sill

77

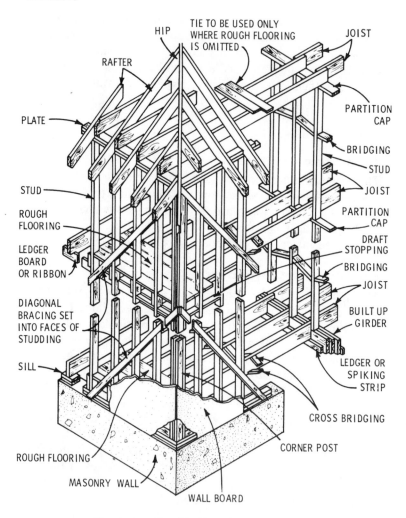

Fig. 1. Details of balloon frame construction.

and second floor joist framing, and sills used in plank and beam framing, and in Fig. 6 is shown the platform construction.

Size of Sills

The size of sills for small buildings of light frame construction consists of 2″ × 6″ lumber, which has been found to be large

Fig. 2. Plank and beam construction.

enough under most conditions. For two-story buildings, and especially in locations subject to earthquakes or tornadoes, a double sill is desirable, as it affords a larger nailing surface for diagonal sheathing brought down over the sill, and ties the wall framing more firmly to its sills. In cases where the building is supported by posts or piers, it is necessary to increase the sill size, since the sill supported by posts acts as a girder. In balloon framing, for example, it is customary to build up the sills with two or more planks 2 or 3 inches thick, which are nailed together.

In most types of construction, since it is not necessary that the sill be of great strength, the foundation will provide uniform solid bearing throughout its entire length. The main requirements are: resistance to crushing across the grain; ability to withstand decay and attacks of insects; and to furnish adequate nailing area for studs, joists, and sheathing.

Length of Sill

The length of the sill is determined by the size of the building, and the foundation should be laid out accordingly. Dimension

79

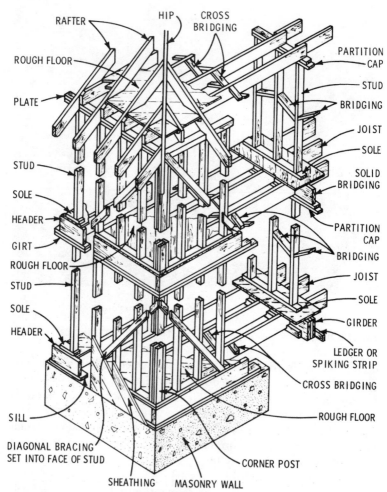

RAFTER **HIP** **CROSS BRIDGING**

ROUGH FLOOR

PLATE

STUD

SOLE

HEADER

GIRT

ROUGH FLOOR

STUD

SOLE

HEADER

SILL

DIAGONAL BRACING SET INTO FACE OF STUD

SHEATHING **MASONRY WALL**

PARTITION CAP

STUD

BRIDGING

JOIST

SOLE

SOLID BRIDGING

PARTITION CAP

BRIDGING

JOIST

SOLE

GIRDER

LEDGER OR SPIKING STRIP

CROSS BRIDGING

ROUGH FLOOR

CORNER POST

Fig. 3. Details of platform construction.

lines for the outside of the building are generally figured from the outside face of the subsiding or sheathing, which is about the same as the outside finish of unsheathed buildings.

Anchorage of Sill

It is important, especially in locations of strong winds, that buildings be thoroughly anchored to the foundation. Anchoring

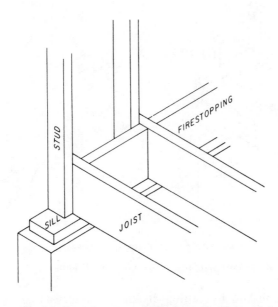

Fig. 4. Details of balloon framing of sill plates. Courtesy of the National Forest Products Assn.

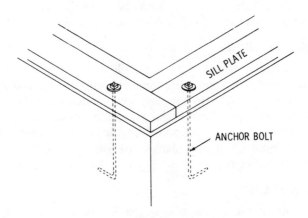

Fig. 5. One way to anchor sill to foundation and joist framing on second floor of braced framing. Courtesy of the National Forest Products Assn.

81

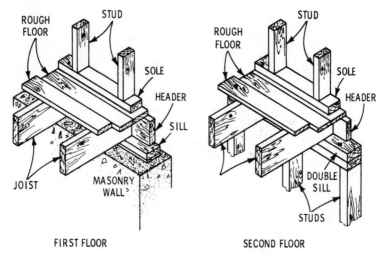

FIRST FLOOR · SECOND FLOOR

BOX-SILL CONSTRUCTION WESTERN FRAME

Fig. 6. Details of platform framing of sill plates and joists.

is accomplished by setting at suitable intervals (6 to 8 feet) ½-inch bolts that extend at least 18 inches into the foundation. They should project above the sill to receive a good size washer and nut. With hollow tile, concrete blocks, and material of cellular structure, particular care should be taken in filling the cells in which the bolts are placed solidly with concrete.

Splicing of Sill

As previously stated, a 2″ × 6″ sill is large enough for small buildings under normal conditions, if properly bedded on the foundations. In order to properly accomplish the splicing of a sill, it is necessary that special precaution be taken. A poorly fitted joint weakens rather than strengthens the sill frame. Where the sill is built up of two planks, the joints in the two courses should be staggered.

Placing of Sill

It is absolutely essential that the foundation be level when placing the sill. It is considered good practice to spread a bed of mortar on the top foundation blocks and lay the sill upon it at once, tapping it gently to secure an even bearing throughout its length. The nuts of the anchoring bolts can then be put in place over the washers and tightened lightly. The nut is securely tightened only after the mortar has had time to harden. In this way a good bearing of the sill is provided, which prevents air leakage between the sill and the foundation wall.

GIRDERS

A girder in small house construction consists of a large beam at the first-story line, which takes the place of an interior foundation wall and supports the inner ends of the floor joists. In a building where the space between the outside walls is more than 14 to 15 feet, it is generally necessary to provide additional support near the center to avoid the necessity of excessively heavy floor joists. When a determination is made as to the number of girders and their location, consideration should be given to the required length of the joists, to the room arrangement, and to the location of the bearing partitions.

LENGTH OF JOISTS

In ordinary cases one girder will generally be sufficient, but if the joist span exceeds 15 feet, a $2'' \times 10''$ joist is usually required, making necessary another girder.

Bridging Between Joists

Cross bridging consists of diagonal pieces (usually $1'' \times 3''$ or $1'' \times 4''$) formed in an X pattern and arranged in rows running at right angles to the joists, as shown in Fig. 7. Their function is to provide a means of stiffening the floors and to distribute the floor load equally. Each piece should be nailed with two or three nails at the top, with the lower ends left loose until the subflooring is in place.

Solid bridging is sometimes used in place of cross bridging. Solid bridging serves the same function as cross bridging, but is made of solid lumber the size of the joist, Fig. 8.

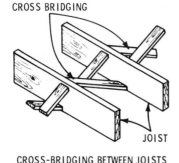

CROSS BRIDGING

JOIST

CROSS-BRIDGING BETWEEN JOISTS

Fig. 7. Cross bridging between floor joists.

Fig. 8. Solid bridging between joists.

INTERIOR PARTITIONS

An interior partition differs from an outside partition in that it seldom rests on a solid wall. Its supports therefore require careful consideration, making sure they are large enough to carry the required weight. The various interior partitions may be bearing or nonbearing, and may run at either right angles or parallel to the joists upon which they rest.

Partitions Parallel to Joists

Here the entire weight of the partition will be concentrated on one or two joists, which perhaps are already carrying their full share of the floor load. In most cases, additional strength should be provided. One method is to provide double joists under such partitions—to put an extra joist beside the regular ones. Computation shows that the average partition weighs nearly three times as much as a single joist should be expected to carry. The usual (and approved method) is to double the joists under nonbearing partitions. An alternative method is to place a joist on each side of the partition.

Partitions at Right Angles to Joists

For nonbearing partitions, it is not necessary to increase the size or number of the joists. The partitions themselves may be braced, but even without bracing, they have some degree of rigidity.

PORCHES

All porches not resting directly upon the ground are normally supported by various posts or piers. Unless porches are entirely enclosed, and of two or more stories in height, their weight will not require a massive foundation.

In pier foundations used to carry concentrated loads, the following sizes will be sufficient:

For very small porches, such as stoops 4 to 6 feet square, concrete footings 12 inches square and 6 inches thick should be used.

For the usual type of front or side porch, the supports of which are not over 10 feet apart, concrete footings 18 inches square and 8 inches thick should be used.

For large porches (especially if they are enclosed), the piers of which exceed a 10-foot spacing, footings 21 or 24 inches square and 10 or 12 inches thick should be used.

Large enclosed porches, especially if more than one story, should have footings and foundations similar to those of the main building.

Porch Joists

Porch joists differ from floor joists, because of their need for greater weather resisting qualities due to their exposed location. In order that rain water may run off freely, the porch should slope approximately ⅛ inch per foot away from the wall of the building. Since there is no subfloor, this requires the flooring to run in the direction of the slope (at right angles to the wall). Otherwise, water will stand on the floor wherever there are slight irregularities.

This construction requires an arrangement comprising a series of girders running from the wall to the piers. The joists run at right angles to these girders and rest on top, or are cut in between them. To protect the tops of the joists from rot, place a strip of 45-pound roofing on the top edge under the flooring.

FRAMING AROUND OPENINGS

It is necessary that some parts of the studs be cut out around windows or doors in outside walls or partitions. It is imperative to insert some form of a header to support the lower ends of the top studs that have been cut off. There is a member that is termed a rough sill at the bottom of the window openings. This sill serves as a nailer, but does not support any weight.

HEADERS

Headers are of two classes, namely:

1. Nonbearing headers, which occur in walls that are parallel with the joists of the floor above and carry only the weight of the framing immediately above.
2. Load-bearing headers, in walls which carry the end of the floor joists either on plates or rib bands immediately above the openings and must therefore support the weight of the floor or the floors above.

Size of Headers

The determining factor in header sizes is whether they are load bearing or not. In general, it is considered good practice to use a double 2 × 4 header placed on edge unless the opening in a nonbearing partition is more than 3 feet wide. In cases where the trim inside and outside is too wide to prevent satisfactory nailing over the openings, it may become necessary to double the header to provide a nailing base for the trim.

CORNER STUDS

These are studs that occur at the intersection of two walls at right angles to each other. A satisfactory arrangement of corner studs is shown in Fig. 9.

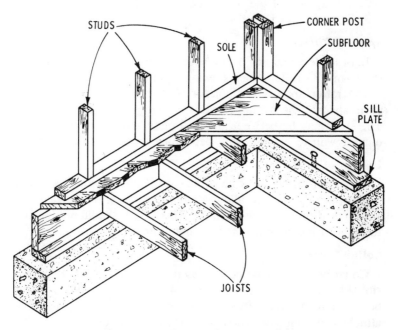

Fig. 9. Detail view of a corner stud.

ROOFS

Generally, it may be said that rafters serve the same purpose for a roof as joists do for floors. They provide a support for sheathing and roof material. Among the various kinds of rafters used, regular rafters extending without interruption from the eave to the ridge are the most common.

Spacing of Rafters

Spacing of rafters is determined by the stiffness of the sheathing between rafters, by the weight of the roof, and by the rafter span. In most cases, the rafters are spaced 16 or 24 inches on center.

87

Size of Rafters

The size of the rafters will depend upon the following factors:

1. The span.
2. The weight of the roof material.
3. The snow and wind loads.

Span of Rafters

In order to avoid any misunderstanding as to what is meant by the span of a rafter, the following definition is given:

The rafter span is the horizontal distance between the supports and not the overall length from end to end of the rafter. The span may be between the wall plate and the ridge, or from outside wall to outside wall, depending on the type of roof.

Length of Rafters

Length of rafters must be sufficient to allow for the necessary cut at the ridge and to allow for the protection of the eaves as determined by the drawings used. This length should not be confused with the span as used for determining length.

Collar Beams

Collar beams may be defined as ties between rafters on opposite sides of a roof. If the attic is to be utilized for rooms, collar beams may be treated as ceiling rafters, with ceiling material attached to them. In general, collar beams should not be relied upon as ties. The closer the ties are to the top, the greater the leverage action. There is a tendency for the collar beam nails to pull out, and also for the rafter to bend if the collar beams are too low. The function of the collar beam is to stiffen the roof. These beams are often, although not always, placed at every rafter. Placing them at every second or third rafter is usually sufficient.

Size of Collar Beams

If the function of a collar beam were merely to resist a thrust, it would be unnecessary to use material thicker than 1-inch lumber. But, as stiffening the roof is their real purpose, the beams must have sufficient body to resist buckling, or else must be braced to prevent bending.

HIP RAFTERS

The hip roof is built in such a shape that its geometrical form is that of a pyramid, sloping down on all four sides. A hip is formed where two adjacent sides meet. The hip rafter is the one that runs from the corner of the building upward along the same plane of the common rafter. Where hip rafters are short and the upper ends come together at the corner of the roof (lending each other support), they may safely be of the same size as that of the regular rafters.

For longer spans, however, and particularly when the upper end of the hip rafter is supported vertically from below, an increase in size is necessary. The hip rafter will necessarily be slightly wider than the jacks, in order to give sufficient nailing surface. A properly sheathed hip roof is nearly self-supporting. It is the strongest roof of any type of framing in common use.

DORMERS

The term dormer is given to any window protruding from a roof. The general purpose of a dormer may be to provide light or to add to the architectural effect.

In general construction, there are three types of dormers, as follows:

1. Dormers with flat sloping roofs, but with less slope than the roof in which they are located, as shown in Fig. 10.
2. Dormers with roofs of the gable type at right angles to the roof.
3. A combination of the above types, which gives the hip-type dormer (Fig. 11).

When framing the roof for a dormer window, an opening is provided in which the dormer is later built in. As the spans are usually short, light material may be used.

STAIRWAYS

The well-built stairway is something more than a convenient means of getting from one floor to another. It must be placed in

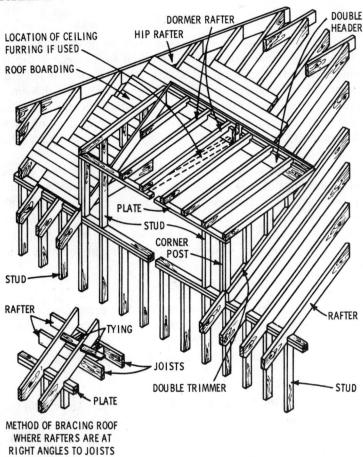

LOCATION OF CEILING FURRING IF USED

ROOF BOARDING

HIP RAFTER

DORMER RAFTER

DOUBLE HEADER

PLATE

STUD

CORNER POST

STUD

RAFTER

TYING

RAFTER

JOISTS

DOUBLE TRIMMER

STUD

PLATE

METHOD OF BRACING ROOF WHERE RAFTERS ARE AT RIGHT ANGLES TO JOISTS

Fig. 10. Detail view of a flat roof dormer.

the right location in the house. The stairs themselves must be so designed that traveling up or down can be accomplished with the least amount of discomfort.

The various terms used in stairway building are as follows:

1. The rise of a stairway is the height from the top of the lower floor to the top of the upper floor.
2. The run of the stairs is the length of the floor space occupied by the construction.

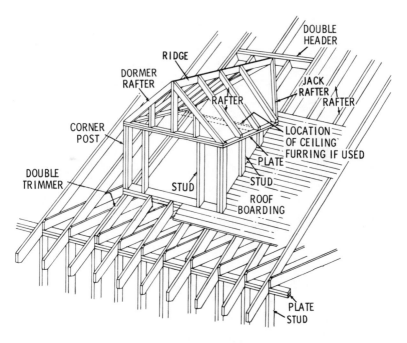

Fig. 11. Detail view of a hip roof dormer.

3. The pitch is the angle of inclination at which the stairs run.
4. The tread is that part of the horizontal surface on which the foot is placed.
5. The riser is the vertical board under the front edge of the tread.
6. The stringers are the frames on the sides which are cut to support the steps.

A commonly followed rule in stair construction is that the tread should not measure less than 9 inches, and the riser should not be more than 8 inches high, as shown in Fig. 12. The width measurement of the tread and height of the riser combined should not exceed 17 inches. Measurements are the cuts of the strings, not the actual width of the boards used for risers and treads. Treads usually have a projection, called a nosing, beyond the edge of the riser.

91

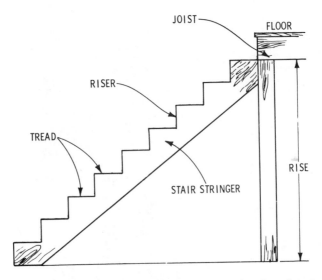

Fig. 12. A method of laying out stair stringers showing the rise and tread.

FIRE AND DRAFT STOPS

Many fires start on lower floors. It is therefore important that fire stops be provided in order to prevent a fire from spreading through the building by way of air passages between the studs. Similarly, fire stops should be provided at each floor level to prevent flames from spreading through the walls and partitions from one floor to the next. Solid blocking should be provided between joists and studs to prevent fire from passing across the building.

In the platform frame and plank and beam framing, the construction itself provides stops at all levels. In these types of construction, therefore, fire stops are needed only in the floor space over the bearing partitions. Masonry is sometimes utilized for fire stopping, but is usually adaptable in only a few places. Generally, obstructions in the air passages may be made of 2-inch lumber, which will effectively prevent the rapid spread of fire. Precautions should be made to ensure the proper fitting of fire stops throughout the building.

CHIMNEY AND FIREPLACE CONSTRUCTION

Although the carpenter is ordinarily not concerned with the building of the chimney, it is necessary that he be acquainted with the methods of framing around it.

The following minimum requirements are recommended:

1. No wooden beams, joists, or rafters should be placed within 2 inches of the outside face of a chimney. No woodwork should be placed within 4 inches of the back wall of any fireplace.
2. No studs, furring, lathing, or plugging should be placed against any chimney or in the joints thereof. Wooden construction should either be set away from the chimney or the plastering should be directly on the masonry or on metal lathing or on incombustible furring material.
3. The walls of fireplaces should never be less than 8 inches thick if brick, or 12 inches if stone.

SUMMARY

Lumber is usually milled into standard lengths, widths, and thicknesses. It is important that the actual rather than just nominal dimensions be taken into consideration when planning a structure. Lumber shipped from the sawmill is divided into three main classes, namely, yard lumber, structural lumber, and factory or shop lumber. Yard lumber is used for house framing and is identified by grades.

Sill plates rest upon the foundation walls and may be anchored with bolts embedded in the foundation wall at approximately 6- to 8-foot intervals. These bolts should project above the sill plate to receive a large-size washer and nut. The sill plate is usually placed on a fresh bed of mortar and tapped gently in place to secure an even bearing seat throughout the length of the sill.

Framing for a building consists of such items as posts, girders, beams, joists, subfloor, sole plates, studs, and rafters. Each particular phase of framing must meet with good material and workmanship to withstand such forces as tornadoes and earth-

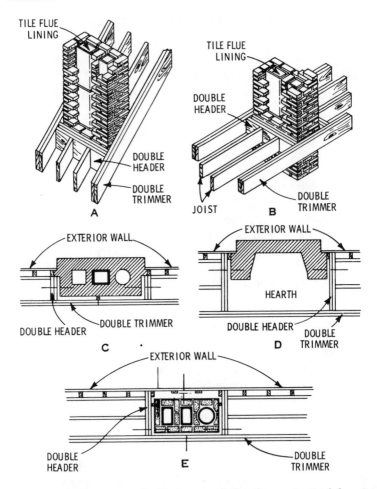

Fig. 13. Framing around chimneys and fireplaces; A. Roof framing around chimney; B. Floor framing around chimney; C. Framing around chimney above fireplace; D. Floor framing around fireplace; E. Framing around concealed chimney above fireplace.

quakes, and control of cracks due to settlement of the foundation. Balloon, plank and beam, and platform framing are three distinct methods of construction.

REVIEW QUESTIONS

1. Name the three types of light frame construction.
2. How are sill plates anchored to the foundation walls?
3. What items are included in framing a building?
4. How are board feet computed?
5. What is cross bridging between floor joists, and why is it so important?

CHAPTER 6

Framing Terms

SILLS

The sills are the first part of the framing to be set in place. Sills may rest either directly on foundation piers or other type of foundation, and usually extend all around the building. Where double or built-up sills are used, the joints are staggered and the corner joints are lapped, as shown in Fig. 1. When box sills are used, that part of the sill that rests on the foundation wall is called the sill plate.

GIRDERS

A girder may be either a single beam or a composite section. Girders usually support joists, whereas the girders themselves

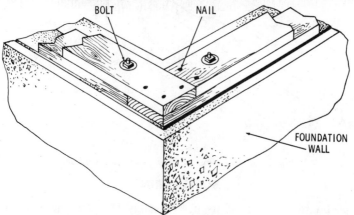

Fig. 1. Details of a double or built-up sill construction.

are supported by bearing walls or columns. When a girder is supported by a wall or pier, it must be remembered that such a member delivers a large concentrated load to a small section of the wall or pier. Therefore, care must be taken to see that the wall or pier is strong enough to carry the load imposed upon it by the girder. Girders are generally used only where the joist will not safely span the distance. The size of a girder is determined by the span length and the load to be carried.

JOISTS

Joists are the pieces that make up the body of the floor frame, and to which the flooring and subflooring are nailed (see Fig. 2). They are usually 2 or 3 inches thick with a varying depth to suit conditions. Joists are usually considered to carry a uniform load composed of the weight of the joists themselves, in addition to the flooring and the weight of furniture and persons.

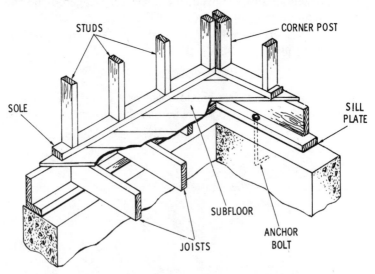

Fig. 2. Detail view of floor joists and subfloor.

SUBFLOORING

A board subfloor, if used, is usually laid diagonally on the joists and properly nailed to them, as shown in Fig. 3. By the

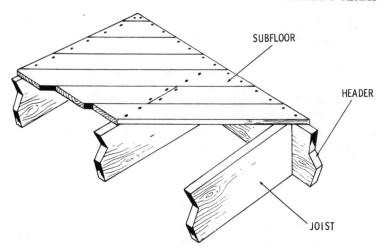

Fig. 3. Method of laying wooden subflooring.

use of subflooring, floors are made much stronger since the floor weight is distributed over a larger area. It may be laid before or after the walls are framed, but preferably before. The subfloor can then be used as a floor to work on while framing the walls. The material for subflooring can be $1'' \times 6''$ sheathing boards or plywood.

HEADERS AND TRIMMERS

A girder is often necessary when an opening is to be made in a floor. The timbers on each side of such an opening are called trimmers, and these must be made heavier than ordinary joists. A piece called a header must be framed in between the trimmers to receive the ends of the short joists.

WALLS

All walls and partitions in which the structural elements are wood are classed as frame construction. Their structural elements are usually closely spaced, and contain a number of slender vertical members termed studs. These are arranged in a row with their ends bearing on a long longitudinal member called the bottom plate or sole plate, and their tops capped with another plate called the top plate. The bearing strength of the stud walls is governed by the length of the studs.

CORNER POSTS

After the sill and first-floor joists are in place, the corner posts are set up, plumbed, and temporarily braced. The corner post may be made in several different ways, depending upon the size and construction of the building.

LEDGER PLATES

In connecting joists to girders and sills where piers are used, a $2'' \times 4''$ piece is nailed to the face of the sill or girder and flush with the bottom edge, as shown in Fig. 4. This is called a ledger.

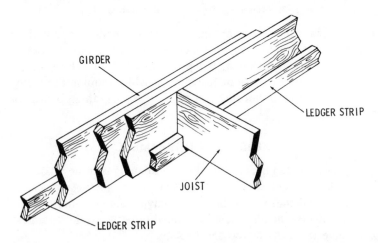

GIRDER

LEDGER STRIP

JOIST

LEDGER STRIP

Fig. 4. Detail of girder and method of supporting joists by using a ledger strip.

TOP PLATES AND SOLE PLATES

The top plate serves two purposes—to tie the studding together at the top and form a finish for the walls, and to furnish a support for the lower end of the rafters. The top plate further serves as a connecting link between the ceiling and the walls.

The plate is made up of one or two pieces of timber of the

same size as the studs. In cases where studs at the end of the building extend to the rafters, no plate is used. When it is used on top of partition walls, it is sometimes called a cap. The sole plate is the bottom horizontal member on which the studs rest.

BRACES

Braces are used as a permanent part of the structure. They tend to stiffen the walls and to keep corners square and plumb. Braces also prevent the frame from being distorted by lateral forces, such as wind or by settlement. These braces are placed wherever the sills or plates make an angle with the corner post, or with a T-post in the outside wall. The brace extends from the sill or sole plate to the top of the post, forming an angle of approximately 60 degrees with the sole plate and an angle of 30 degrees with the post.

STUDS

Studs are the main vertical framing members. They are installed between the top and sole plates. Before studs are laid out, the windows and door openings are laid out. The studs are normally set 16 inches apart (on centers) but there is a trend today toward 24-inch centers. In most instances, the studs are nailed to the top and sill plates while the entire assembly is resting flat on the deck. The wall is then raised into position.

PARTITION WALLS

Partition walls are any walls that divide the inside space of a building. In most cases, these walls are framed as a part of the building. In cases where floors are to be installed after the outside of the building is completed, the partition walls are left unframed. There are two types of partition walls, the bearing and the nonbearing. The bearing type, also called load bearing, supports the structure above. The nonbearing type supports only itself, and may be put in at any time after the framework is installed. Only one cap or plate is used. Some craftsmen refer to nonbearing walls as "partition type."

A sole plate should be used in every case as it helps to distribute the load over a large area. Partition walls are framed in the same manner as outside walls, and inside door openings are framed the same as outside openings. Where there are corners (or where one partition wall joins another), corner posts or T-posts are used in the outside walls. These posts provide nailing surfaces for the inside wall finish.

BRIDGING

Walls of frame buildings are bridged in most cases to increase their strength. There are two methods of bridging, the diagonal and the horizontal. Diagonal bridging is nailed between the studs at an angle. It is more effective than horizontal bridging because it forms a continuous truss and tends to keep the walls from sagging. Whenever possible, both outside and inside walls should be bridged alike.

Horizontal bridging is nailed between the studs and halfway between the sole and top plate. This bridging is cut to lengths that correspond to the distance between the studs at the bottom. Such bridging not only stiffens the wall but will also help straighten the studs.

Metal bridgings are forged, thin strips of metal secured to tops of joists.

RAFTERS

In all roofs, the pieces that make up the main body of the framework are called the rafters. They do for the roof what the joists do for the floor or the studs do for the wall. The rafters are inclined members usually spaced from 16 to 24 inches on center, which rest at the bottom on the plate and are fastened on center at the top in various ways according to the form of the roof. The plate forms the connecting link between the wall and the roof and is really a part of both.

The size of the rafters varies, depending upon the length and the distance at which they are spaced. The connection between the rafter and the wall is the same in all types of roofs. They may or may not extend out a short distance from the wall to form the eaves and to protect the sides of the building.

SUMMARY

After the foundation walls have been completed, the first part of framing is to set the sills in place. The sills usually extend all around the building. Where double or built-up sills are used, the joints are staggered and the corner joints are lapped.

Where floor joists span a large distance, girders are used to help support the load. The size of a girder depends on the span length of the joists. Girders may be either a steel I-beam or can be built on the job from 2 × 8 or 2 × 10 lumber placed side by side.

Joists are the pieces that make up the body of the floor frame, and to which the subflooring is nailed. The joists are usually 2 × 6 or 2 × 8 lumber commonly spaced 16 or 24 inches apart. The subflooring is generally installed before the wall construction is started, and in this manner the subfloor can serve as a work floor.

REVIEW QUESTIONS

1. Explain the purpose of the sill plate, floor joists, and girder.
2. What is the purpose of the subflooring?
3. What is a ledger strip?
4. What is a sole plate?
5. Where are headers and trimmers used?

CHAPTER 7

Girders and Sills

Chapter 5 presented a variety of framing techniques and sections. Here, a number of these things, starting with girders and sills, are explained in detail, showing the variety of framing options available. As with any work involving carpentry, an overall idea obtains: Use quality materials. In many if not most cases building codes specify the wood that must be used, but sometimes codes do not specify wood, and inferior substitutions are made. This is poor economy, and if the inferior material is in a structurally crucial position, it could be unsafe.

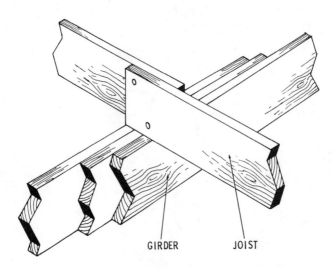

GIRDER   JOIST

Fig. 1. Illustration of girder and joist construction.

GIRDERS

By definition, a girder is a principal beam extending from wall to wall of a building affording support for the joists or floor beams where the distance is too great for a single span. Girders may be either solid or built up, as shown in Figs. 1 and 2.

Construction of Girders

Girders can be built up of wood if select stock is used. Be sure it is straight and sound. If the girders are to be built up of 2 × 8 or 2 × 10 stock, place the pieces on the sawhorses and nail them together. Use the piece of stock that has the least amount of warp for the center piece and nail other pieces on the sides of the center stock. Use a common nail that will go through the first piece and nearly through the center piece. Square off the ends of the girder after the pieces have been nailed together. If the stock is not long enough to build up the girder the entire length, the pieces must be built up by staggering the joints. If the girder supporting post is to be built up, it is to be done in the same manner as described for the girder.

Table 1 gives the size of built-up wood girders for various loads and spans, based on Douglas fir 4-square framing lumber.

All girders are figured as being made with 2-inch dressed stock. The 6-inch girder is figured three pieces thick; the 8-inch

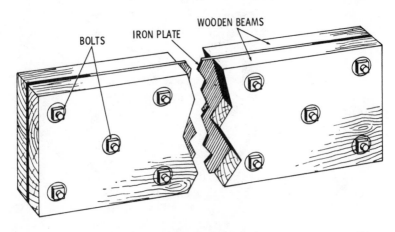

Fig. 2. A flitch plate girder.

girder four pieces thick; the 10-inch girder with five pieces thick, and the 12-inch girder with six pieces thick. For solid girders, multiply the load in Table 1 by 1.130 when 6-inch girders are used; 1.150 when 8-inch girders are used; 1.170 when 10-inch girders are used, and 1.180 when 12-inch girders are used.

Table 1. Nominal Size of Girder Required

Load per linear foot of girder	Length of Span				
	6-foot	7-foot	8-foot	9-foot	10-foot
750	6x8 in	6x8 in	6x8 in	6x10 in	6x10 in
900	6x8	6x8	6x10	6x10	8x10
1050	6x8	6x10	8x10	8x10	8x12
1200	6x10	8x10	8x10	8x10	8x12
1350	6x10	8x10	8x10	8x12	10x12
1500	8x10	8x10	8x12	10x12	10x12
1650	8x10	8x12	10x12	10x12	10x14
1800	8x10	8x12	10x12	10x12	10x14
1950	8x12	10x12	10x12	10x14	12x14
2100	8x12	10x12	10x14	12x14	12x14
2250	10x12	10x12	10x14	12x14	12x14
2400	10x12	10x14	10x14	12x14	x
2550	10x12	10x14	12x14	12x14	x
2700	10x12	10x14	12x14	x	x
2850	10x14	12x14	12x14	x	x
3000	10x14	12x14	x	x	x
3150	10x14	12x14	x	x	x
3300	12x14	12x14	x	x	x

Placing Basement Girders

Basement girders must be lifted into place on top of the piers, and walls built for them, and set perfectly level and straight from end to end. Some carpenters prefer to give the girders a slight crown of approximately 1 inch in the entire length, which is a wise plan because the piers generally settle more than the outside walls. When there are posts instead of brick piers used to support the girder, the best method is to temporarily support the girder by uprights made of 2×4 joists resting on blocks on the ground below. When the superstructure is raised, these can be knocked out after the permanent posts are placed. The practice of temporarily shoring the girders, and not placing the permanent supports until after the superstructure is finished, is favored by many builders, and it is a good idea for carpenters to know just how it should be done. Permanent supports are usually made by using 3- or 4-inch steel pipe set in a concrete foundation, or footing that is at least 12 inches deep. They are called Lally columns.

SILLS

A sill is that part of the side walls of a house that rests horizontally upon, and is securely fastened to, the foundation. There are various types of sills.

Sills may be divided into two general classes:

1. Solid.
2. Built up.

The built-up sill has become more or less a necessity because of the scarcity and high cost of timber, especially in the larger sizes. The work involved in sill construction is very important for the carpenter. The foundation wall is the support upon which the entire structure rests. The sill is the foundation on which all the framing structure rests, and is the real point of departure for actual carpentry and joinery activities. The sills are the first part of the frame to be set in place. They either rest directly on the foundation piers or on the ground, and may extend all around the building.

The type of sill used depends upon the general type of construction.

Sills are called:

1. Box sills.
2. T-sills.
3. Braced framing sills.
4. Built-up sills.

Box Sills — Box sills are often used with the very common style of platform framing, either with or without the sill plate. In this type of sill the part that lies on the foundation wall or ground is called the sill plate. The sill is laid edgewise on the outside edge of the sill plate. It is doubled and bolted to the foundation. Aluminum is set in asphaltum beneath it as a termite barrier.

T-Sills—There are two types of T-sill construction, one commonly used in the South or in a dry warm climate, and one used in the North and East where it is colder. Their construction is similar except in the East the T-sills and joists are nailed directly to the studs, as well as to the sills. The headers are nailed in between the floor joists.

Braced Framing Sills—The floor joists are notched out and nailed directly to the sills and studs.

Built-up Sills—Where built-up sills are used, the joists are staggered. If piers are used in the foundation, heavier sills are used. These sills are of single heavy timber or are built up of two or more pieces of timber. Where heavy timber or built-up sills are used, the joints should occur over the piers. The size of the sill depends on the load to be carried and on the spacing of the piers. Where earth floors are used, the studs are nailed directly to the sill plates.

Setting the Sills

After the girder is in position, the sills are placed on top of the foundation walls, are fitted together at the joints, and leveled throughout. The last operation can be done either by a sight level or by laying them in a full bed of mortar and leveling them with the anchor bolts, as shown in Fig. 3. Or it can be left "loose" and then all the bolts can be tightened as needed to bring the sill level.

Sills that are to rest on a wall of masonry should be kept up at least 18 inches above the ground, as decaying sills are a frightful source of trouble and expense in wooden buildings. Sheathing, paper, and siding should therefore be very carefully installed to exclude all wind and wet weather.

SUMMARY

A girder is a principal beam extending from wall to wall of a building to support the joists or floor beams where the distance is too great for a single span. Girders can be made of wood, which must be straight and free of knots. Girders are made three, four, five, or six pieces thick, depending on the load per linear foot and length of the girder.

109

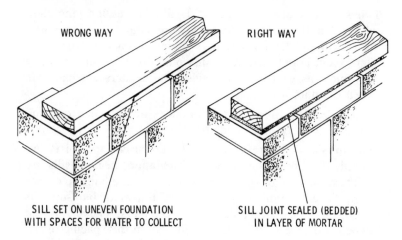

Fig. 3. Wrong and right way to set sills. The top solid block serves as a termite barrier.

A sill is that part of the side walls of a house that rests horizontally upon, and is securely fastened to, the foundation. The sills are the first part of the framing to be set in place; therefore, it is important that the sill be constructed properly and placed properly on the foundation. There are various types of sills used; which sill depends on the type of house construction. The size of the sill depends on the load to be carried and on the spacing of the piers. In some construction, two sill plates are used, one nailed on top of the other.

REVIEW QUESTIONS

1. Why are girders used?
2. How are sills anchored to the foundation?
3. Name the various types of sills constructed.
4. What is the purpose of the sill?
5. What is the purpose of the sill?

CHAPTER 8

Floor Framing

After the girders and sills have been placed, the next operation consists in sawing to size the floor beams or joists of the first floor, and placing them in position on the sills and girders. If there is a great variation in the size of timbers, it is necessary to cut the joists ½ inch narrower than the timber so that their upper edges will be in alignment. This sizing should be made from the top edge of the joist, as shown in Fig. 1. When the joists have been cut to the correct dimension they should be placed upon the sill and girders, and spaced 16 inches between centers, beginning at one side or end of a room. This is done to avoid wasting the material.

CONNECTING JOIST TO SILLS AND GIRDERS

Joists can be connected to sills and girders by several methods, but the prime consideration, of course, is to be sure

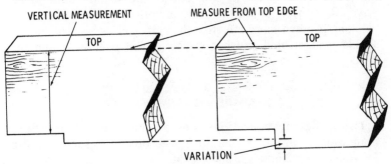

VERTICAL MEASUREMENT   MEASURE FROM TOP EDGE

TOP   TOP

VARIATION

Fig. 1. Showing the variation of joist width.

that the connection is able to hold the load that the joists will carry.

The placing of the girders is an important factor in making the connection. The joists must be level; therefore, if the girder is not the same height as the sill, the joists must be notched. In placing joists, always have the crown up, since this counteracts the weight on them; in most cases there will be no sag below a straight line. When a joist is to rest on plates or girders, the joist is cut long enough to extend the full width of the plate or girder.

BRIDGING

To prevent joists from springing sideways under load, which would reduce their carrying capacity, they are tied together diagonally by 1 × 3 or 2 × 3 strips, a process called *bridging*. The 1 × 3 ties are used for small houses and the 2 × 3 stock on larger work. Metal bridging may also be used (Fig. 2).

Rows of bridging should not be more than 8 feet apart. Bridging pieces may be cut all in one operation with a miter box, or the bridging may be cut to fit. Bridging is put in before the subfloor is laid, and each piece is fastened with two nails at the top end. The subfloor should be laid before the bottom end is nailed.

A more rigid (less vibrating) floor can be made by cutting in solid 2-inch joists of the same depth. They should be cut perfectly square and a little full, say $1/16$ inch longer than the inside distance between the joists. First, set a block in every other space, then go back and put in the intervening ones. This keeps the joists from spreading and allows the second ones to be driven in with the strain the same in both directions. This solid blocking is much more effective than cross-bridging. The blocks should be toenailed and not staggered and nailed through the joists.

HEADERS AND TRIMMERS

The foregoing operations would complete the first-floor framework in rooms having no framed openings, such as those for stairways, chimneys, and elevators.

The definition of a header is a short transverse joist that sup-

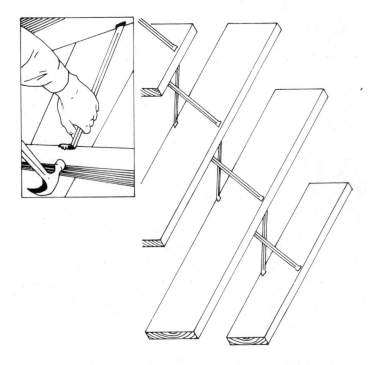

Fig. 2. Metal bridging in place and being installed.

ports the ends of one or more joists (called tail beams) where they are cut off at an opening. A trimmer is a carrying joist which supports an end of a header.

Shown in Fig. 3 are typical floor openings used for chimneys and stairways. For these openings, the headers and trimmers are set in place first, then the floor joists are installed. Hanger irons of the type shown in Fig. 4 are nearly obsolete, but many other kinds of metal hangers are available, such as the one shown in Fig. 5.

Headers run at right angles to the direction of the joists and are doubled. Trimmers run parallel to the joists and are actually doubled joists. The joists are framed to the headers where the headers form the opening frame at right angles to the joists. These shorter joists framed to the headers are called *tail beams*.

113

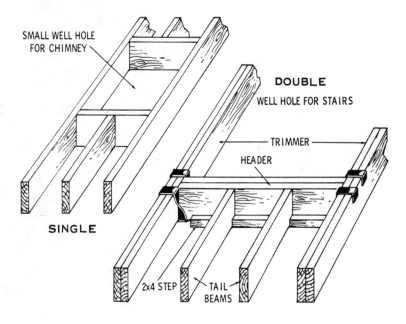

SMALL WELL HOLE
FOR CHIMNEY

DOUBLE
WELL HOLE FOR STAIRS

TRIMMER

HEADER

SINGLE

2x4 STEP

TAIL
BEAMS

Fig. 3. Single and double header and trimmer construction for floor openings.

The number of headers and trimmers required at any opening depends upon the shape of the opening.

SUBFLOORING

With the sills and floor joists completed, it is necessary to install the subflooring. The subflooring is permanently laid before erecting any wall framework, since the wall plate rests on it. In most cases the subfloor is laid diagonally, as shown in Fig. 6, to give strength and to prevent squeaks in the floor. This floor is called the rough floor, or subfloor, or deck, and may be viewed as a large platform covering the entire width and length of the building. Two layers or coverings of flooring material (subflooring and finished flooring) are placed on the joists. The subfloor may be 1″ × 4″ square edge stock, 1″ × 6″ or 1″ × 8″ shiplap, or 4′ × 8′ sheets of plywood.

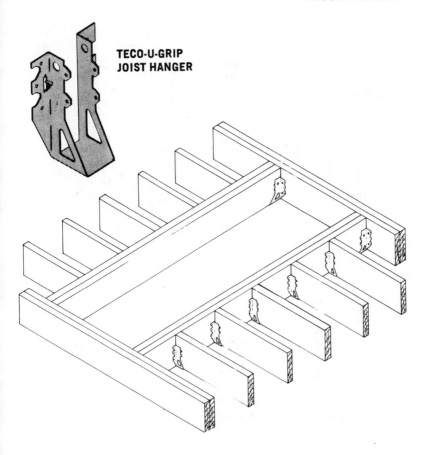

TECO-U-GRIP JOIST HANGER

Fig. 4. One of many framing fasteners available. These are joist hangers but are commonly referred to as "Tecos," after the brand name.

FINISH FLOORING

A finished wood floor is in most cases ¾-inch tongue-and-groove hardwood, such as oak. Prefinished flooring can also be obtained. In addition, there are nonwood materials, which will be covered in detail later.

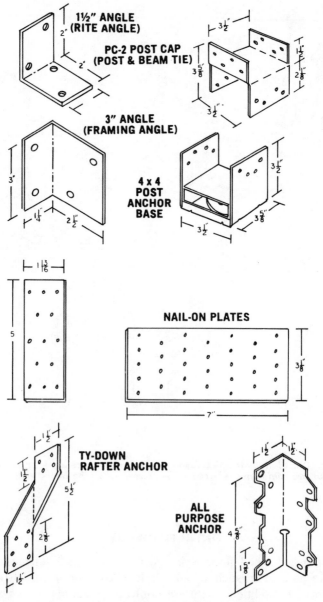

Fig. 5. Other fasteners available from Teco.

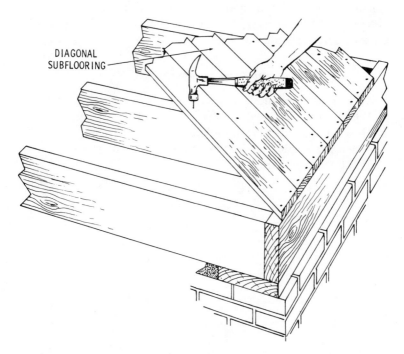

DIAGONAL
SUBFLOORING

Fig. 6. Board subflooring is laid diagonally for greater strength.

SUMMARY

After the sill plate and girder have been constructed and installed, the next operation is to install the floor joists. The joists can be installed in a variety of ways.

To prevent joists from warping under load, which would reduce their carrying capacity, they are tied together diagonally with 1×3 or 2×3 boards or metal bridging. Bridging is installed before the subflooring is laid. The subfloor should be laid before the bottom end is nailed.

After the joists are installed, the subflooring is laid. In most cases, board subfloor is laid diagonally to give strength and prevent squeaks in the floor. The material generally used for subfloors is 1×6 or 1×8 boards, or $4' \times 8'$ plywood panels.

The finished floor in most cases is ¾-inch tongue-and-groove hardwood. To save time and labor, a prefinished flooring can be obtained. Where heavy loads are to be carried on the floor, 2-inch flooring should be used.

REVIEW QUESTIONS

1. What must be done to align the top edge of floor joists?
2. What is bridging, and how is it installed?
3. How should subflooring be laid?
4. When are headers and trimmers used in floor joists?
5. What are tail beams?

CHAPTER 9

Outer Wall Framing

Where the construction permits, it is advisable to lay the sub-floor permanently before erecting the wall studding. Besides saving time in laying and removing a temporary floor, this procedure also furnishes a surface on which to work and a sheltered place in the basement for storage of tools and materials.

BUILT-UP CORNER POSTS

There are many ways in which corner posts may be built up, using studding or larger sized pieces. Some carpenters form corner posts with two 2 × 4 studs spiked together to make a piece having a 4 × 4 section. Except for very small structures, such a flimsily built-up piece should not be used as a corner post. Fig. 1 shows various arrangements of built-up posts commonly used.

BALLOON BRACING

In balloon framing, the bracing may be temporary or permanent. Temporary bracing means strips are nailed on (as in Fig. 2) to hold the frame during construction and are removed when the permanent bracing is in place. There are two kinds of permanent bracing. One type is called *cut-in* bracing and is shown in Fig. 3. A house braced in this manner withstood one of the worst hurricanes ever to hit the eastern seaboard, and engineers, after an inspection, gave the bracing the full credit for the survival. The other type of bracing is called *plank* bracing and is shown in Fig. 4. This type is put on from the outside, the studs being cut and

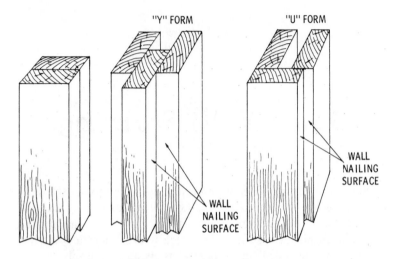

Fig. 1. Various ways to build up corner post.

notched so that the bracing is flush with the outside edges of the studs. This method of bracing is commonly used and is very effective.

PREPARING THE CORNER POSTS AND STUDDING

In laying out the posts and studs, a pattern should be used to ensure that all will be of the same length, with the gains, and any other notches or mortises to be cut, at the same elevation. The pattern can be made from a ¾-inch board, or a 2 × 4 stud, and must be cut to the exact length and squared at both ends. The pattern should be made from a selected piece of wood having straight edges.

If a stud is selected for the pattern, it may be used later in the building and thus counted in with the total number of pieces to be framed. The wall studs are placed on their edges on the horses in quantities of 6 to 10 at a time and marked from the pattern, as shown in Fig. 5.

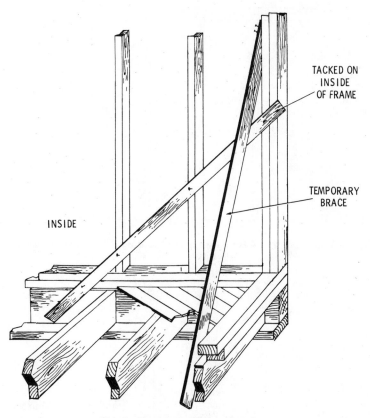

TACKED ON
INSIDE
OF FRAME

TEMPORARY
BRACE

INSIDE

Fig. 2. Temporary frame bracing.

ERECTING THE FRAME

If the builder is short-handed, or working by himself, a frame can be *crippled* together, one member at a time. An experienced carpenter can erect the studs and toenail them in place with no assistance whatever. Then the corners are plumbed and the top plates nailed on from a stepladder. Where enough men are available, most contractors prefer to nail the sole plates and top plates as the entire wall is lying down on the rough floor. Some even cut all the door and window openings, after which the entire gang raises the assembly and nails it in position. Needless to say, the platform frame with the subfloor in place is best

121

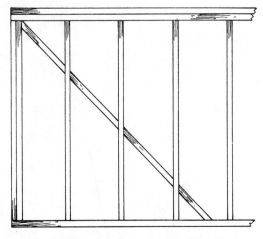

Fig. 3. Permanent 2 × 4 cut between stud bracing. This type of bracing is called cut-in braces.

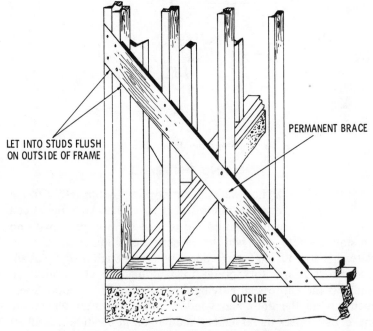

LET INTO STUDS FLUSH
ON OUTSIDE OF FRAME

PERMANENT BRACE

OUTSIDE

Fig. 4. Permanent plank-type bracing.

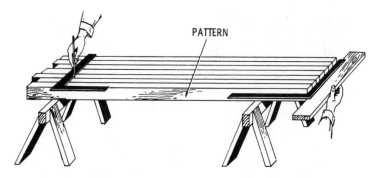

Fig. 5. Marking wall studs from pattern.

adapted to this kind of construction. It is probably the speediest of any possible method, but it cannot be done by only one or two men. Some contractors even put on the outside sheathing before raising the wall, and then use a lift truck or a highlift excavating machine to raise it into position.

There are many ways to frame the openings for windows and doors. Some builders erect all of the wall studding first, and then cut out the studs for the openings. It is much easier to saw studs lying level on horses than when nailed upright in the frame. The openings should be laid out and framed complete. Studding at all openings must be double to furnish more area for proper nailing of the trim, and must be plumb, as shown in Fig. 6.

OPENING SIZES FOR WINDOWS AND DOORS

Standard double-hung windows are listed by unit dimensions, giving the width first. A 2'-0'' × 3'-0'' double hung window has an overall glass size of 28½ inches. The rough opening is 2'-10'', and the sash opening is 2'-8''.

It is common to frame the openings with dimensions 2 inches more than the unit dimension. This will allow for the plumbing and leveling of the window.

For door openings, allow a full 1 inch over the door size for the thickness of the jambs, with about 1 inch additional for blocking and plumbing. This will make the opening width 3 inches over the door size. Standard oak thresholds are ⅝ inch

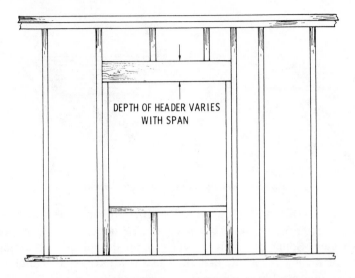

DEPTH OF HEADER VARIES
WITH SPAN

Fig. 6. An approved framing for window opening.

thick, making the overall height of the door-frame butt ⅝ inch above the finish floor line. If you wish to allow for the lugs on the frame (which is a good idea), allow 5 inches over the door size for the height of the opening above the finish floor line. Door casings are set to show a slight rabbet, or set back, to allow for clearance of the hinges, while windows are usually cased flush with the inside edges of the jambs. If the heights of the window and door head casings are to be the same (as they often are), the window frames must be set the height of this rabbet *higher* than the door frames. Continuous head trim is almost a thing of the past but where it is to be used, the carpenter can easily get into an almost uncorrectable situation if this difference is neglected.

PARTITIONS

Interior walls that divide the inside space of the buildings into rooms or halls are known as partitions. These are made up of studding covered with plasterboard and plaster, metal lath and plaster, or dry wall. Fig. 7 shows the studding of an ordinary partition, each joist being 16 inches on center. Where partitions

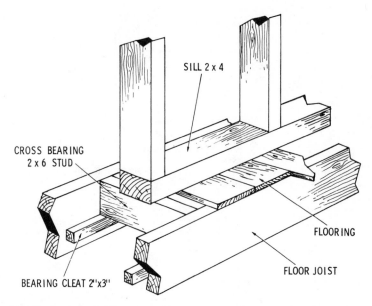

Fig. 7. A method of constructing a partition between two floor joists.

are placed between and parallel to floor joist, bridges must be placed between the joists to provide a means of fastening the partition plate.

The construction at the top is shown in Fig. 8. Openings over 30 inches wide in partitions or outside walls must have heavy headers, as shown in Fig. 9. The partition wall studs are arranged in a row with their ends bearing on a long horizontal member called a bottom plate, and their tops capped with another plate, called a top plate. Double top plates are used in bearing walls and partitions. The bearing strength of stud walls is determined by the strength of the studs.

Walls and partition coverings are divided into two general types—wet-wall material (generally plaster), and drywall material (also called plasterboard, sheetrock, and gypsum board). Drywall material comes in 4' × 8' and 4' × 12' sheets, and in various thicknesses from ³/₈ inch up; ³/₈, ¹/₂, and ⁵/₈ inch are the most common sizes. It is normally applied in single thickness. When covering both walls and ceiling, it is best to start at the ceiling.

125

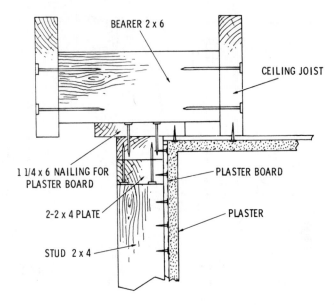

Fig. 8. A method of constructing a partition between two ceiling joists.

You can build two T-braces to hold the panels in place, as shown in Fig. 10. Raise the first panel in position and hold it in place with the T-brackets. Each panel should be nailed at least every 4 inches, starting at the center of the board and working outward. Due to the weight of drywall-boards, a special nail can be purchased that will not pull out and let the ceiling sag. Dimple the nail below the surface of the panel, being careful not to break the surface of the board by the blow of the hammer.

SHEETROCKING WALLS

In general, you should install drywall from the top down. This reduces the amount of taping to be done; you will have short joints instead of long ones.

We favor a horizontal application of drywall because the joints are at waist height rather than higher up. Mark stud locations on the ceiling before covering walls; you will know where they are for nailing.

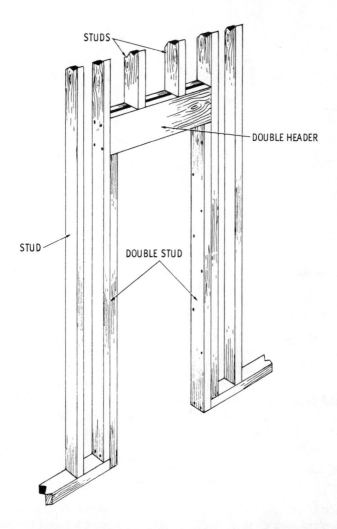

STUDS

DOUBLE HEADER

STUD

DOUBLE STUD

Fig. 9. Construction of headers when openings over 30 inches between studs appear in partitions or outside walls.

First, pick up a panel, hike it into position, then tack it in place to hold it there. Now nail it home. Put in the panel below it in the same way. (You will, of course, have to cut the panels to fit.)

127

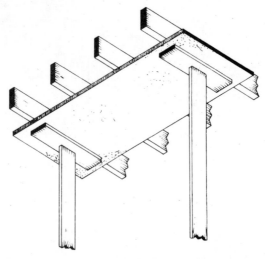

Fig. 10. T-braces used to temporarily support sheets of drywall on the ceiling.

TAPING

To tape, use a broad-edged knife. First apply a coat about 16 inches wide. Into this set the tape, smoothing it in place with the knife. When dry, apply a 12-inch-wide coat, let it dry, then finish with an 18-inch coat. Also use the joint cement over nailheads. Together, all coats should be no more than $1/16$ inch thick.

A good taping job should not require sanding. But if it does, sand it after priming. The paint will reveal any fuzziness that must be removed.

SUMMARY

There are generally two types of permanent bracing, cut-in and plank, which is a very important part of outer wall framing.

If a carpenter is working alone, he can erect each stud and toenail into the sole plate with no assistance whatsoever. The top plate can be nailed from a stepladder, after the corners and studs are plumb. When enough men are available, most contractors prefer to nail the sole plates and top plates as the entire wall is lying on the floor or ground. In many cases, all of the door and

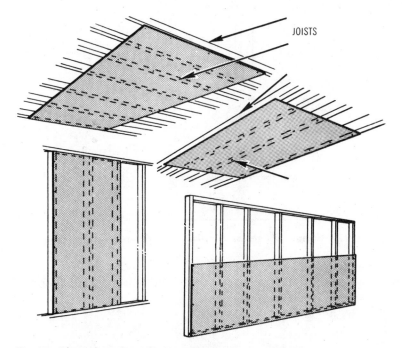

JOISTS

Fig. 11. Plasterboard may be installed across or with joists, and vertically or horizontally on studs. Courtesy of 3M Co.

window openings are constructed and nailed in place, then the wall is raised as one unit and nailed in position.

Interior walls that divide the inside space into rooms or halls are generally known as partitions. They are made up of studding covered with plaster or plasterboard, with each joist set 16 inches on center. Where partitions are to be placed between and parallel to the floor joists, bridges must be placed between the joists to provide a means of fastening the partition plate.

REVIEW QUESTIONS

1. Why is corner bracing so important?
2. What is the difference between a load-bearing and a non-load-bearing partition?

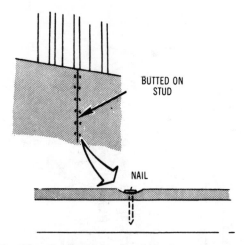

BUTTED ON
STUD

NAIL

Fig. 12. All nails should be dimpled, then covered with compound.

3. What is a bearing member?
4. Why should double headers and studs be installed in door and window openings?
5. Why are corner post designs so important?

CHAPTER 10

Roof Framing

As a preliminary to the study of this chapter, the reader should review Chapter 23 in Vol. 1 on "How To Use the Steel Square." This tool, which is invaluable to the carpenter in roof framing, has been explained in detail in that chapter with many examples of rafter cutting. Hence, a knowledge of how to use the square is assumed here to avoid repetition.

TYPES OF ROOFS

There are numerous forms of roofs and a seemingly endless variety of shapes. The carpenter and the student, as well as the architect, should be familiar with the names and features of each of the various types.

Shed or Lean-to Roof

This is the simplest form of roof (shown in Fig. 1) and is usually employed for small sheds and outbuildings. It has a single slope and is not a thing of beauty.

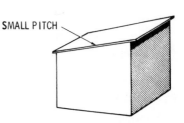

Fig. 1. Shed or lean-to roof used on small sheds or buildings.

SMALL PITCH

Saw-Tooth Roof

This is a development of the shed or lean-to roof, being virtually a series of lean-to roofs covering one building, as in Fig. 2. It is used on factories, principally because of the extra light that can be obtained through windows on the vertical sides.

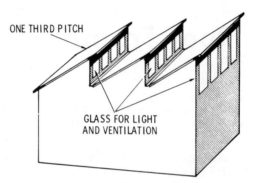

ONE THIRD PITCH

GLASS FOR LIGHT AND VENTILATION

Fig. 2. A sawtooth roof used on factories for light and ventilation.

Gable or Pitch Roof

This is a very common, simple, and efficient form of roof and is used extensively on all kinds of buildings. It is of triangular section, having two slopes meeting at the center or ridge and forming a gable, as in Fig. 3. It is popular because of the ease of construction, relative economy, and efficiency.

Gambrel Roof

This is a modification of the gable roof, each side having two slopes, as shown in Fig. 4.

Hip Roof

A hip roof is formed by four straight sides, all sloping toward the center of the building, and terminating in a ridge instead of a deck, as in Fig. 5.

Pyramid Roof

A modification of the hip roof in which the four straight sides sloping toward the center terminate in a point instead of a ridge,

Fig. 3. Gable or pitch roof can be used on all types of buildings. Courtesy of Shetter-Kit, Inc.

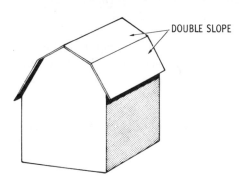

DOUBLE SLOPE

Fig. 4. Gambrel roof used on barns.

as in Fig. 6. The pitch of the roof on the sides and ends is different. This construction is not often used.

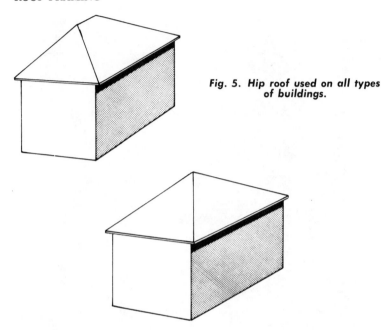

Fig. 5. Hip roof used on all types of buildings.

Fig. 6. Pyramid roof which is not often used.

Hip-and-Valley Roof

This is a combination of a hip roof and an intersecting gable called because both hip and valley rafters are required in its construction. There are many modifications of this roof. Usually the intersection is at right angles, but it need not be; either ridge may rise above the other and the pitches may be equal or different, thus giving rise to an endless variety, as indicated in Fig. 7.

Double-Gable Roof

This is a modification of a gable or a hip-and-valley roof in which the extension has two gables formed at its end, making an M-shape section, as in Fig. 8.

Ogee Roof

A pyramidal form of roof having steep sides sloping to the center, each side being ogee-shaped, lying in a compound hollow and round curve, as in Fig. 9.

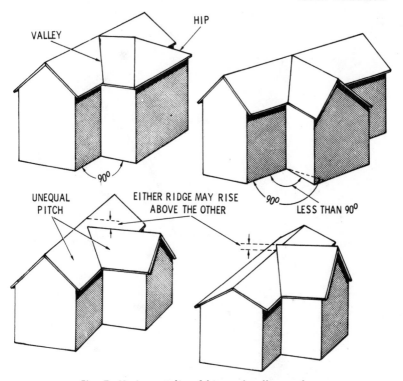

Fig. 7. Various styles of hip and valley roofs.

Mansard Roof

The straight sides of this roof slope vary steeply from each side of the building toward the center, and the roof has a nearly flat deck on top, as in Fig. 10. It was introduced by the architect whose name it bears.

French or Concave Mansard Roof

This is a modification of the Mansard roof, its sides being concave instead of straight, as in Fig. 11.

Conical Roof or Spire

A steep roof of circular section that tapers uniformly from a circular base to a central point. It is frequently used on towers, as in Fig. 12.

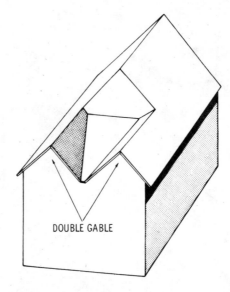

Fig. 8. Double gable roof.

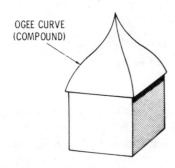

Fig. 9. Ogee style roof.

Dome

A hemispherical form of roof (Fig. 13) used chiefly on observatories.

ROOF CONSTRUCTION

The frame of most roofs is made up of timbers called rafters. These are inclined upward in pairs, their lower ends resting on

136

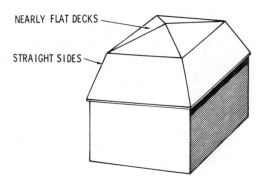

NEARLY FLAT DECKS

STRAIGHT SIDES

Fig. 10. Mansard roof.

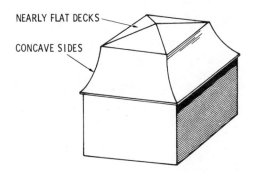

NEARLY FLAT DECKS

CONCAVE SIDES

Fig. 11. French or concave Mansard roof.

the top plate and their upper ends being tied together with a ridge board. On large buildings, such framework is usually reinforced by interior supports to avoid using abnormally large timbers.

One prime object of a roof in any climate is to keep out water. The roof must be sloped or otherwise built to shed water. Where heavy snows cover the roof for long periods of time, it must be constructed more rigidly to bear the extra weight. Roofs must also be strong enough to withstand high winds.

The most commonly used types of roof construction include:

1. Gable.
2. Lean-to or shed.

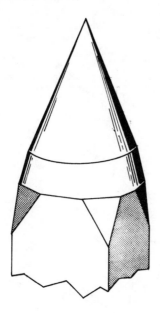

Fig. 12. Conical or spire roof.

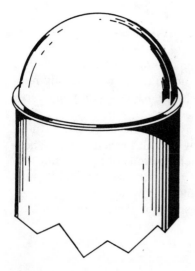

Fig. 13. Dome-type roof.

3. Hip.

4. Gable and valley.

Terms used in connection with roofs are span, total rise, total run, unit of run, rise in inches, pitch, cut of roof, line length, and plumb and level lines.

Span—The span of any roof is the shortest distance between the two opposite rafter seats. Stated another way, it is the measurement between the outside plates, measured at right angles to the direction of the ridge of the building.

Total Rise—The total rise is the vertical distance from the plate to the top of the ridge.

Total Run—The term total run always refers to the level distance over which any rafter passes. For the ordinary rafter, this would be one-half the span distance.

Unit of Run—The unit of measurement, 1 foot or 12 inches, is the same for the roof as for any other part of the building. By the use of this common unit of measurement, the framing square is employed in laying out large roofs.

Rise in Inches—The rise in inches is the number of inches that a roof rises for every foot of run.

Pitch—Pitch is the term used to describe the amount of slope of a roof.

Cut of Roof—The cut of a roof is the rise in inches and the unit of run (12 inches).

Line Length—The term line length as applied to roof framing is the hypotenuse of a triangle whose base is the total run and whose altitude is the total rise.

Plumb and Level Lines—These terms have reference to the direction of a line on a rafter and not to any particular rafter cut. Any line that is vertical when the rafter is in its proper position is called a plumb line. Any line that is level when the rafter is in its proper position is called a level line.

RAFTERS

Rafters are the supports for the roof covering and serve in the same capacity as joists do for the floor or studs do for the walls.

According to the expanse of the building, rafters vary in size from ordinary 2 × 4s to 2 × 10s. For ordinary dwellings, 2 × 6 rafters are used, spaced from 16 to 24 inches on centers.

The various kinds of rafters used in roof construction are:

1. Common.
2. Hip.
3. Valley.
4. Jack (hip, valley, or cripple).
5. Octagon.

The carpenter should thoroughly know these various types of rafters, and be able to distinguish each kind as they are briefly described.

Common Rafters

A rafter extending at right angles from plate to ridge, as shown in Fig. 14.

Hip Rafter

A rafter extending diagonally from a corner of the plate to the ridge, as shown in Fig. 15.

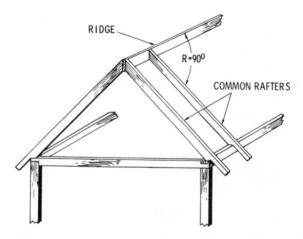

Fig. 14. Common rafters.

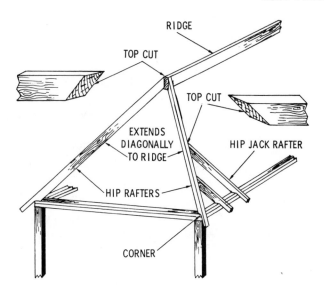

Fig. 15. Hip roof rafters.

Valley Rafter

A rafter extending diagonally from the plate to the ridge at the intersection of a gable extension and the main roof.

Jack Rafter

Any rafter which does not extend from the plate to the ridge.

Hip Jack Rafter

A rafter extending from the plate to a hip rafter, and at an angle of 90° to the plate, as shown in Fig. 15.

Valley Jack Rafter

A rafter extending from a valley rafter to the ridge and at an angle of 90° to the ridge, as shown in Fig. 16.

Cripple Jack Rafter

A rafter extending from a valley rafter to hip rafter and at an angle of 90° to the ridge, as shown in Fig. 17.

141

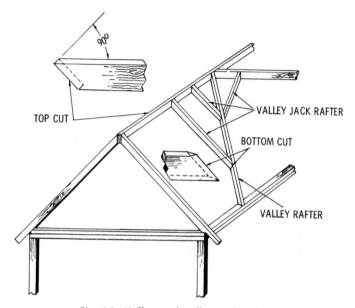

Fig. 16. Valley and valley jack rafters.

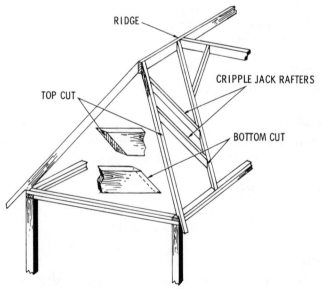

Fig. 17. Cripple jack rafters.

Octagon Rafter

Any rafter extending from an octagon-shaped plate to a central apex, or ridge pole.

A rafter usually consists of a main part or rafter proper, and a short length called the *tail,* which extends beyond the plate. The rafter and its tail may be all in one piece, or the tail may be a separate piece nailed onto the rafter.

LENGTH OF RAFTER

The length of a rafter may be found in several ways:

1. By calculation.
2. With steel framing square.
   a. Multi-position method.
   b. By scaling.
   c. By aid of the framing table.

Example—What is the length of a common rafter having a run of 6 feet and rise of 4 inches per foot?

1. *By calculation* (See Fig. 18)

The total rise = 6 × 4 = 24 inches = 2 feet.

Since the edge of the rafter forms the hypotenuse of a right triangle whose other two sides are the run and rise, then the length of the rafter = $\sqrt{\text{run}^2 + \text{rise}^2} = \sqrt{6^2 + 2^2} = \sqrt{40} =$ 6.33 feet, as illustrated in Fig. 18.

Practical carpenters would not consider it economical to find rafter lengths in this way because it takes too much time and there is a chance of error. It is to avoid both these objections that the *framing square* has been developed.

2. *With steel framing square*

The steel framing square considerably reduces the mental effort and chances of error in finding rafter lengths. An approved method of finding rafter lengths with the square is by the aid of

143

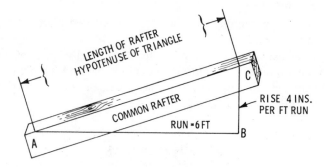

Fig. 18. Method of finding the length of a rafter
by calculation.

the rafter table included on the square for that purpose. However, some carpenters may possess a square which does not have rafter tables. In such case, the rafter length can be found either by the *multi-position* method shown in Fig. 19, or by *scaling* as in Fig. 20. In either of these methods, the measurements should be made with care because, in the *multi-position* method, a separate measurement must be made for each foot run with a chance for error in each measurement.

Problem: Lay off the length of a common rafter having a run of 6 ft. and a rise of 4 in. per ft. Locate a point A

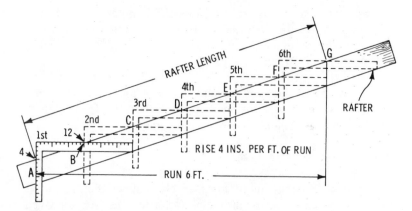

Fig. 19. Multi-position method of finding rafter length.

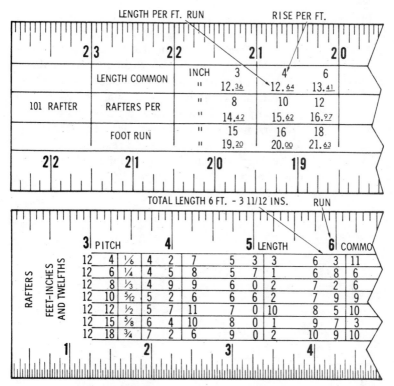

Fig. 20. Rafter table readings of two well-known makers of steel framing squares.

on the edge of the rafter, leaving enough stock for a lookout, if any is used. Place the steel framing square so that division 4 coincides with *A*, and 12 registers with the edge of *B*. Evidently, if the run were 1 ft., distance *AB* thus obtained would be the length of the rafter *per foot run*. Apply the square six times for the 6-ft. run, obtaining points *C, D, E, F,* and *G*. The distance *AG*, then, is the length of the rafter for a given run.

Fig. 20 shows readings of rafter tables of two well-known makes of squares for the length of the rafter in the preceding example, one giving the length per foot run, and the other the total length for the given run.

Problem 1: *Given the rise per ft. in inches.* Use two squares, or a square and a straightedge scale, as shown in Fig. 21. Place the straightedge on the square so as to be able to read the length of the diagonal between the rise of 4 in. on the tongue and the 1-ft. (12-in.) run on the body as shown. The reading is a little over 12 inches. To find the fraction, place dividers on 12 and at point *A,* as in Fig. 22. Transfer to the hundredths scale and read .65, as in Fig. 23, making the length of the rafter 12.65 in. *per ft. run,* which for a 6-ft. run = $\dfrac{12.65 \times 6}{12}$ = 6.33 ft.

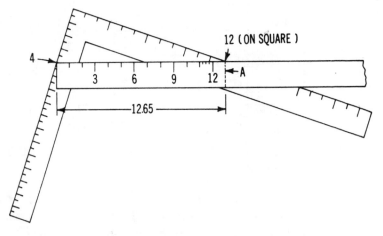

Fig. 21. Method of finding rafter length by scaling.

Problem 2: *Total rise and run given in feet.* Let each inch on the tongue and body of the square = 1 ft. The straightedge should be divided into inches and 12ths of an inch so that on a scale, 1 in. = 1 ft. Each division will therefore equal 1 in. Read the diagonal length between the numbers representing the run and rise (12 and 4), taking the whole number of inches as feet, and the fractions as inches. Transfer the fraction with dividers and

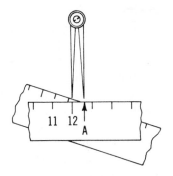

Fig. 22. Reading the straight-edge in combination with the carpenter's square.

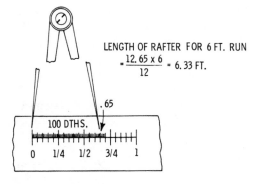

LENGTH OF RAFTER FOR 6 FT. RUN

$$= \frac{12.65 \times 6}{12} = 6.33 \text{ FT.}$$

Fig. 23. Method of reading hundredths scale.

apply the 100th scale, as was done in Problem 1, Figs. 22 and 23.

In estimating the total length of stock for a rafter having a tail, the run of the tail or length of the lookout must of course be considered.

RAFTER CUTS

All rafters must be cut to the proper angle or bevel at the points where they are fastened and, in the case of overhanging rafters, also at the outer end. The various cuts are known as:

1. Top or plumb.
2. Bottom, seat, or heel.
3. Tail or lookout.
4. Side, or cheek.

147

Common Rafter Cuts

All of the cuts for the various types of common rafters are made at right angles to the sides of the rafter; that is, not beveled as in the case of jacks. Fig. 24 shows various common rafters from which the nature of these various cuts is seen.

In laying out cuts from common rafters, one side of the square is always placed on the edge of the stock at 12, as shown in Fig. 24. This distance 12 corresponds to 1 foot of the run; the other side of the square is set with the edge of the stock to the rise in inches *per foot run*. This is virtually a repetition of Fig. 19, but it is very important to understand why one side of the square is set to 12 for common rafters—not simply to know that 12 must be used. On rafters having a full tail, as in Fig. 25B, some carpenters do not cut the rafter tails, but wait until the rafters are set in place so that they may be lined and cut while in position. Certain kinds of work permit the ends to be cut at the same time the remainder of the rafter is framed.

The method of handling the square in laying out the bottom

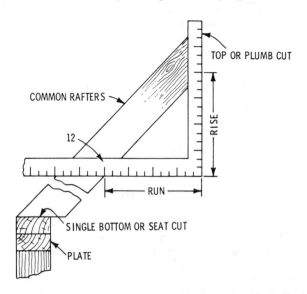

Fig. 24. Various common rafters illustrating types and names of cuts; showing why one side of the square is set at 12 in laying out the cut.

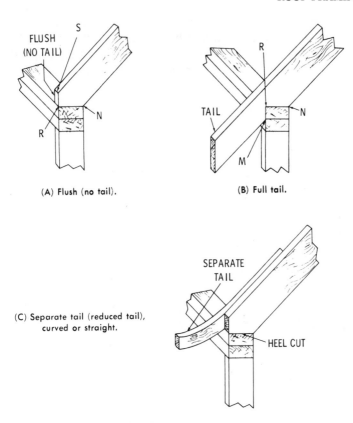

(A) Flush (no tail).

(B) Full tail.

(C) Separate tail (reduced tail),
curved or straight.

Fig. 25. Various forms of common rafter tails.

and lookout cuts is shown in Fig. 26. In laying out the top or plumb cut, if there is a ridge board, one-half of the thickness of the ridge must be deducted from the rafter length. If a lookout or a tail cut is to be vertical, place the square at the end of the stock with the rise and run setting as shown in Fig. 26, and scribe the cut line *LF*. Lay off *FS* equal to the length of the lookout, and move the square up to *S* (with the same setting) and scribe line *MS*. On this line, lay off *MR*, the length of the vertical side of the bottom cut. Now apply the same setting to the bottom edge of the rafter, so that the edge of the square cuts *R*, and scribe *RN*, which is the horizontal side line of the bottom

149

cut. In making the bottom cut, the wood is cut out to the lines *MR* and *RN*. The lookout and bottom cuts are shown made in Fig. 25B, *RN* being the side which rests on the plate, and *RM* the side that touches the outer side of the plate.

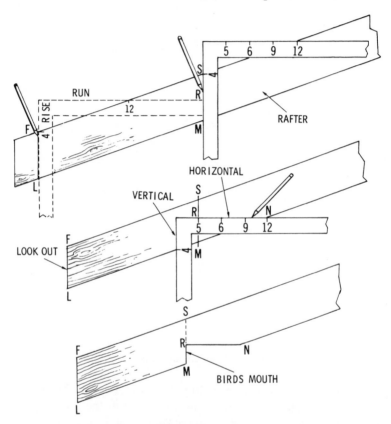

Fig. 26. Method of using the square in laying out the lower or end cut of the rafter.

Hip and Valley Rafter Cuts

The hip rafter lies in the plane of the common rafters and forms the hypotenuse of a triangle, of which one leg is the adjacent common rafter and the other leg is the portion of the plate intercepted between the feet of the hip and common rafters, as in Fig. 27.

150

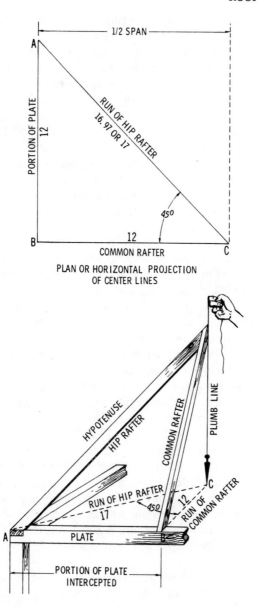

Fig. 27. View of hip and common rafters in respect to each other.

Problem: In Fig. 27, take the run of the common rafter as 12, which may be considered as 1 ft. (12 in.) of the run or the total run of 12 ft. (½ the span). Now for 12 ft., intercept on the plate the hip run inclined to 45° to the common run, as in the triangle ABC. Thus, $AC^2 = \sqrt{AB^2 + BC^2} = \sqrt{12^2 + 12^2} = 16.97$, or approximately 17. Therefore, the run of the hip rafter is to the run of the common rafter as 17 is to 12. Accordingly, in laying out the cuts, use figure 17 on one side of the square and the given rise in *inches per foot* on the other side. This also holds true for top and bottom cuts of the valley rafter when the plate intercept AB = the run BC of the common rafter.

The line of measurement for the length of a hip and valley rafter is along the middle of the back or top edge, as on common and jack rafters. The rise is the same as that of a common rafter, and the run of a hip rafter is the horizontal distance from the plumb line of its rise to the outside of the plate at the foot of the hip rafter, as shown in Fig. 28.

In applying the source for cuts of hip or valley rafters, *use the distance 17 on the body of the square in the same way as 12 was used for common rafters*. When the plate distance between hip and common rafters is equal to half the span or run of the common rafter, the line of run of the hip will lie at 45° to the line of the common rafter, as indicated in Fig. 27.

The length of a hip rafter, as given in the framing table on the square, is the distance from the ridge board to the outer edge of the plate. In practice, deduct from this length one-half the thickness of the ridge board, and add for any projection beyond the plate for the eave. Fig. 29A shows the correction for the table length of a hip rafter to allow for a ridge board, and Fig. 29B shows the correction at the plate end which may or may not be made as in Fig. 30.

The table length, as read from the square, must be reduced an amount equal to MS. This is equal to the hypotenuse (ab) of the little triangle abc, which in value $= \sqrt{ac^2 + bc^2} = \sqrt{ac^2 \times \text{(half thickness of ridge)}^2}$. In ordinary practice, take MS

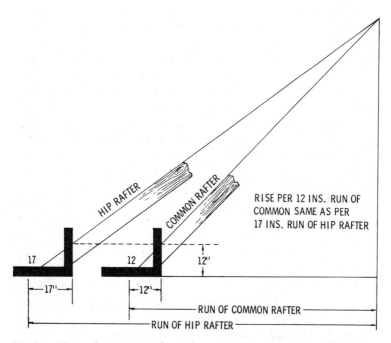

RISE PER 12 INS. RUN OF
COMMON SAME AS PER
17 INS. RUN OF HIP RAFTER

HIP RAFTER

COMMON RAFTER

17

12

12"

17"

12"

RUN OF COMMON RAFTER

RUN OF HIP RAFTER

Fig. 28. Hip and common rafters shown in the same plane, illustrating the use of 12 for the common rafter and 17 for the hip rafter.

as equal to half the thickness of the ridge. The plan and side view of the hip rafter shows the table length and the correction *MS*, which must be deducted from the table length so that the sides of the rafter at the end of the bottom cut will intersect the outside edges of the plate. The table length of the hip rafter, as read on the framing square, will cover the span from the ridge to the outside cover *a* of the plate, but the side edges of the hip intersect the plates at *b* and *c*. The distance that *a* projects beyond a line connecting *bc* or *MS* must be deducted; that is, measured backward toward the ridge end of the hip. In making the bottom cut of a valley rafter, it should be noted that a valley rafter differs from a hip rafter in that the correction distance for the table length must be added instead of subtracted, as for a hip rafter. A distance *MS* was subtracted from the table length of the hip rafter in Fig. 29B, and an equal distance (*LF*) was added for the valley rafter in Fig. 30.

153

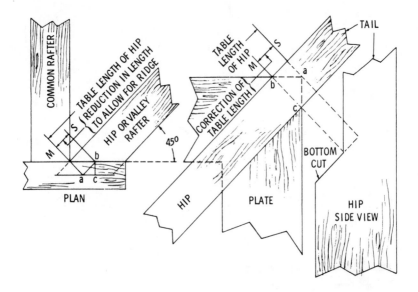

Fig. 29. *Correction in table for top cut to allow for half thickness of ridge board.*

After the plumb cut is made, the end must be mitered outward for a hip, as in Fig. 31, and inward for a valley, as in Fig. 32, to receive the fascia. A fascia is the narrow vertical member fastened to the outside ends of the rafter tails. The miter cuts are shown with full tails in Fig. 33, which illustrates hip and valley rafters in place on the plate.

Side Cuts of Hip and Valley Rafters

These rafters have a side or cheek cut at the ridge end. In the absence of a framing square, a simple method of laying out the side cut for a 45° hip or valley rafter is as follows:

Measure back on the edge of the rafter from point *A* of the top cut, as shown in Fig. 34. Distance *AC* is equal to the thickness of the rafter. Square across from *C* to *B* on the opposite edge, and scribe line *AB*, which gives the side cut. *FA* is the top cut, and *AB* is the side cut. Here *A*, the point from which half the thickness of the rafter is measured, is seen at the top end of the cut.

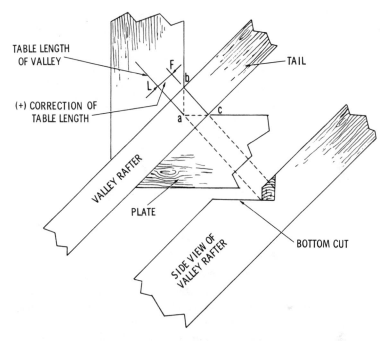

TABLE LENGTH OF VALLEY

(+) CORRECTION OF TABLE LENGTH

TAIL

VALLEY RAFTER

PLATE

SIDE VIEW OF VALLEY RAFTER

BOTTOM CUT

Fig. 30. Side view of valley rafter showing bottom and seat cut at top plate.

This rule does not hold for any angle other than 45°.

BACKING OF HIP RAFTERS

By definition, the term backing is the bevel on the top side of a hip rafter that allows the roofing boards to fit the top of the rafter without leaving a triangular hole between it and the back of the roof covering. The height of the hip rafter, measured on the outside surface vertically upward from the outside corner of the plate, will be the same as that of the common rafter measured from the same line, whether the hip is backed or not. This is not true for an unbacked valley rafter when the measurement is made at the center of the timber.

The graphical method of finding the backing of hip rafters is shown in Fig. 35. Let *AB* be the span of the building, and *OD* and *OC* the plan of two unequal hips. Lay off the given rise as

155

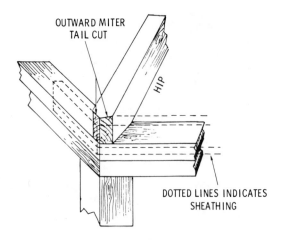

OUTWARD MITER
TAIL CUT

HIP

DOTTED LINES INDICATES
SHEATHING

Fig. 31. Flush hip rafter miter cut.

shown. Then *DE* and *CF* are the lengths of the two unequal hips. Take any point, such as *G* on *DE*, and erect a perpendicular cutting *DF* at *H*. Revolve *GH* to *J*, that is, make *HJ = GH*, draw *NO* perpendicular to *OD* and through *H*. Join *J* to *N* and *O*, giving a bevel angle *NJO*, which is the backing for rafter *DE*. Similarly, the bevel angle *NJO* is found for the backing of rafter *CF*.

JACK RAFTERS

As outlined in the classification, there are several kinds of jack rafters, and they are distinguished by their relation with other rafters of the roof. These various jack rafters are known as:

1. Hip jacks.
2. Valley jacks.
3. Cripple jacks.

The distinction between these three kinds of jack rafters, as shown in Fig. 36, is as follows: Rafters that are framed between a hip rafter and the plate are hip jacks; those framed between the ridge and a valley rafter are valley jacks; those framed between hip and valley rafters are cripple jacks.

The term cripple is applied because the ends or *feet* of the rafters are cut off—the rafter does not extend the full length

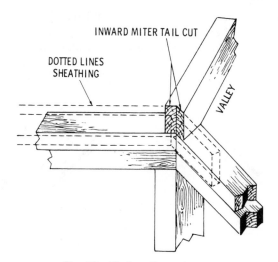

Fig. 32. Flush valley miter cut.

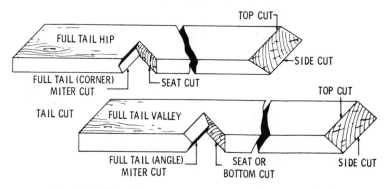

Fig. 33. Full tail hip and valley rafters showing all cuts.

from ridge to plate. From this point of view, a valley jack is sometime erroneously called cripple; it is virtually a semi-cripple rafter, but confusion is avoided by using the term cripple for rafters framed between the hip and valley rafters, as above defined.

Jack rafters are virtually discontinuous common rafters. They are cut off by the intersection of a hip or valley, or both, before reaching the full length from plate to ridge. Their lengths are found in the same way as for common rafters—the number 12

157

being used on one side of the square and the rise in inches per foot run on the other side. This gives the length of jack rafter per foot run, and is true for all jacks—hip, valley, and cripple.

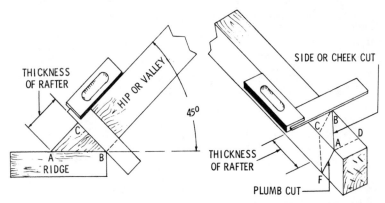

Fig. 34. A method of obtaining a side cut of 45° hip or valley rafter without aid of a framing table.

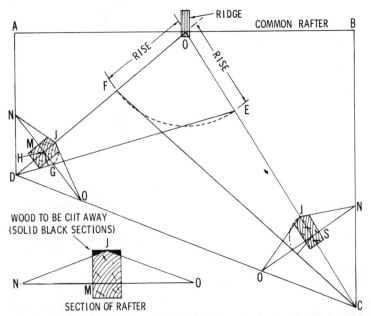

Fig. 35. Graphical method of finding length of rafters and backing of hip rafters.

In actual practice, carpenters usually measure the length of hip or valley jacks from the long point to the ridge, instead of along the center of the top, no reduction being made for one-half the diagonal thickness of the hip or valley rafter. Cripples are measured from long point to long point, no reduction being made for the thickness of the hip or valley rafter.

As no two jacks are of the same length, various methods of procedure are employed in framing, as:

1. Beginning with shortest jack.
2. Beginning with longest jack.
3. Using framing table.

Shortest Jack Method

Begin by finding the length of the shortest jack. Take its spacing from the corner, measured on the plates, which in the case of a 45° hip is equal to the jack's run. The length of this first jack will be the common difference that must be added to each jack to get the length of the next longer jack.

Longest Jack Method

Where the longest jack is a full-length rafter (that is, a common rafter), first find the length of the longest jack, then count the spaces between jacks and divide the length of the longest jack by the number of spaces. The quotient will be the common difference. Then frame the longest jack and make each jack shorter than the preceding jack by this common difference.

Framing Table Method

On various steel squares, there are tables giving the length of the shortest jack rafters corresponding to the various spacings, such as 16, 20, and 24 inches between centers for the different pitches. This length is also the common difference and thus serves for obtaining the length of all the jacks.

Example—Find the length of the shortest jack or the common difference in the length of the jack rafters, where the rise of the roof is 10 inches per foot and the jack rafters are spaced 16 inches between centers; also, when spaced 20 inches between centers. Fig. 37 shows the reading of the jack table on the square

159

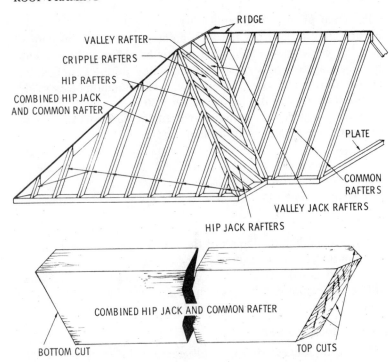

Fig. 36. A perspective view of hip and valley roof showing the various kinds of jack rafters, and enlarged detail of combined hip jack and common rafters showing cuts.

10 IN. RISE PER FT.

LENGTH SHORTEST JACK 16 IN. CENTER

Fig. 37. Square showing table for shortest jack rafter at 16'' on center.

for 16-inch centers, and Fig. 38 shows the reading on the square for 20-inch centers.

Jack-Rafter Cuts

Jack rafters have top and bottom cuts that are laid out the same as for common rafters, and also side cuts that are laid out the same as for a hip rafter. To lay off the top or plumb cut with a square, take 12 on the tongue and the rise in inches (of common rafter) per foot run on the blade, and mark along the blade as in Fig. 39. The following example illustrates the use of the framing square in finding the side cut.

Example—Find the side cut of a jack rafter framed to a 45° hip or valley for a rise of 8 inches per foot run. Fig. 40 shows the reading on the jack side-cut table of the framing square, and Fig. 41 shows the method of placing the square on the timber to ob-

LENGTH SHORTEST JACK · · · · 10 IN. RISE PER FT. RUN

	17		16		15		14	
46	6 18⁰²	DIF IN	INCH ''	3 20⅝	4 21⅛	6 22⅜	DIF IN	
72	12 20⁸⁰	LENGTH OF JACKS	'' ''	8 24	10 26	12 28¼	LENGTH OF JA	
	18 24⁷⁵	20 INCH CENTERS	'' ''	15 32	16 33⅜	18 36	24 INCH CENT	

16 · · · 15 · · · 14 · · · 13 · · · 12

Fig. 38. *Square showing table for shortest jack rafter at 20″ on center.*

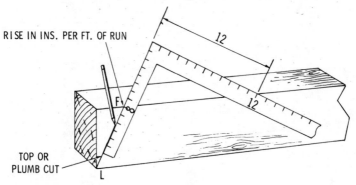

RISE IN INS. PER FT. OF RUN

12

12

F

TOP OR PLUMB CUT

L

Fig. 39. *Method of finding plumb and side cuts of jack framed to 45° hip or valley.*

tain the side cut. It should be noted that different makers of squares use different setting numbers, but the ratios are always the same.

METHOD OF TANGENTS

The tangent value is made use of in determining the side cuts of jack, hip, or valley rafters. By taking a circle with a radius of 12 inches, the value of the tangent can be obtained in terms of the constant of the common rafter run.

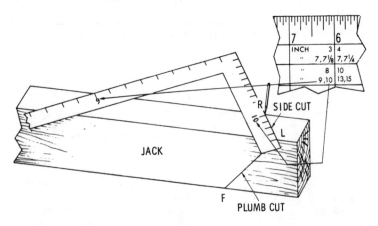

4 25¼	6 26⅞	FIG'S GIVING	INCH "	3 7¾ 8	4 99¼	6 9 10	FIG'S GIVIN
10 31¼	12 34	SIDE CUT	" "	8 10 12	10 10 13	12 12 17	SIDE CUT OF HIP ON
16 40	12 43¼	OF JACKS	" "	15 10 16	16 9 15	18 10 18	VALLEY RAFTER

JACK SIDE CUT   8 IN. RISE PER FT. RUN

Fig. 40. A framing square showing readings for side cut of jack corresponding to 8-inch rise per foot run.

INCH	3 4
"	7,7⅛ 7,7¼
"	8 10
"	9,10 13,15

SIDE CUT

JACK

PLUMB CUT

Fig. 41. Method of placing framing square on jack to lay off side cut for an 8-inch rise.

Considering rafters with zero pitch, as shown in Fig. 42, if the common rafter is 12 feet long, the tangent *MS* of a 45° hip is the same length. Placing the square on the hip, setting to 12 on the tongue and 12 on the body will give the side cut at the ridge when there is no pitch (at *M*), as in Fig. 43. Placing the square on the jack with the same setting numbers (12, 12) as at *S*, will give the face cut for the jack when framed to a 45° hip with zero pitch; that is, when all the timbers lie in the same plane.

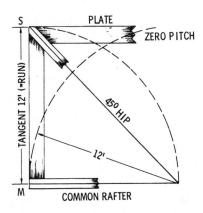

Fig. 42. A roof with zero pitch showing the common rafter and the tangent as being the same length.

OCTAGON RAFTERS

On an octagon, or eight-sided, roof, the rafters joining the corners are called octagon rafters and are a little longer than the common rafter and shorter than the hip or valley rafters of a square building of the same span. The relation between the run of an octagon and a common rafter is shown in Fig. 44 as being as 13 is to 12. That is, for each foot run of a common rafter, an octagon rafter would have a run of 13 inches. Hence, to lay off the top or bottom cut of an octagon rafter, place the square on the timber with the 13 on the tongue and the rise of the common rafter per foot run on the blade, as shown graphically in Fig. 45. The method of laying out the top and bottom cut with the 13 rise setting is shown in Fig. 46.

163

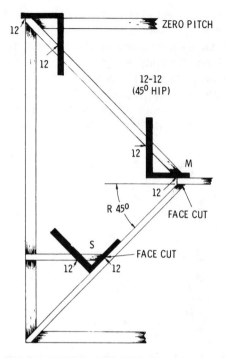

Fig. 43. Zero pitch square 45° roof showing application of the framing square to give side cuts at ridge.

The length of an octagon rafter may be obtained by scaling the diagonal on the square for 13 on the tongue and the rise in inches per foot run of a common rafter, and multiplying by the number of feet run of a common rafter. The principle involved in determining the amount of backing of an octagon rafter (or any other polygon) is the same as for hip rafters. The backing is determined by the tangent of the angle whose adjacent side is ½ the rafter thickness and whose angle is equivalent to one-half the central angle.

TRUSSES

Definite savings in material and labor requirements through the use of preassembled wood roof trusses make truss framing an effective means of cost reduction in small dwelling construc-

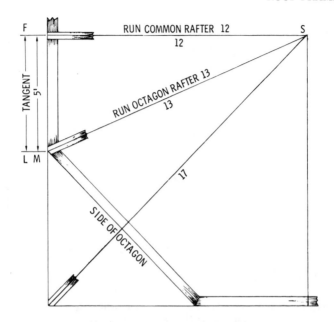

F RUN COMMON RAFTER 12 S
12
TANGENT 5'
RUN OCTAGON RAFTER 13
13
L M
17
SIDE OF OCTAGON

Fig. 44. Details of an octagon roof showing relation
in length between common and octagon rafters.

tion. In a 26' × 32' dwelling, for example, the use of trusses can result in a substantial cost saving and a reduction in use of lumber of almost 30 percent as compared with conventional rafter and joist construction. In addition to cost savings, roof trusses offer other advantages of increased flexibility for interior planning, and added speed and efficiency in site erection procedures. Today, some 70 percent of homes built in the United States use trusses.

For many years, trusses have been used extensively in commercial and industrial buildings, and are very familiar in bridge construction. But there are still pockets of the country that know nothing about them.

During the last several years, however, special efforts have been made to get the truss used in small homes. One of the results has been the development of light wood trusses that permit substantial savings in the case of lumber. Not only may the framing lumber be smaller in dimension than in conventional

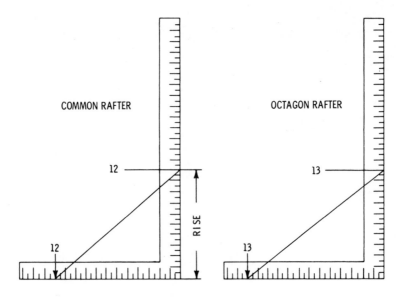

Fig. 45. For equal rise, the run of octagon rafters is 13 inches, to 12 inches for the common rafters.

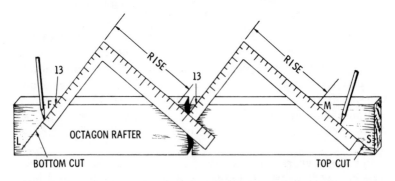

Fig. 46. Method of laying off bottom and top cuts of an octagon rafter with a square using the 13 rise setting.

framing, but trusses may also be spaced 24 inches on center as compared to the usual 16-inch spacing of rafter and joist construction.

The following figures show percentage savings in the 26′ × 32′ house through the use of trusses 24 inches on center:

Lumber Requirements for Trusses

28.4 percent less than for conventional framing at 16-inch spacing.

Labor Requirements for Trusses

36.8 percent less hours than for conventional framing at 16-inch spacing.

Total Cost of Trusses

29.1 percent less than conventional framing at 16-inch spacing.

The trusses consisted of 2 × 4 lumber at top and bottom chords, 1 × 6 braces, and double 1 × 6s for struts, with plywood gussets and splices, as shown in Fig. 47. The clear span of truss construction permits use of nonbearing partitions so that it is possible to eliminate the extra top plate required for bearing partitions used with conventional framing. It also permits a smaller floor girder to be used for floor construction since the floor does not have to support the bearing partition and help carry the roof load.

Aside from direct benefits of reduced cost and savings in material and labor requirements, roof trusses offer special advantages in helping to speed up site erection and overcome delays due to weather conditions. These advantages are reflected not only in improved construction methods but also in further reductions in cost. With preassembled trusses, a roof can be put over the job quickly to provide protection against the weather.

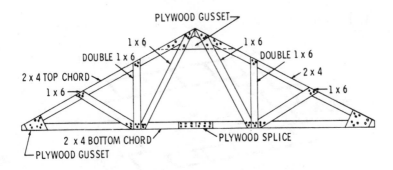

Fig. 47. Wood roof truss for small dwellings.

167

Laboratory tests and field experience show that properly designed roof trusses are definitely acceptable for dwelling construction. The type of truss shown in Fig. 47 is suitable for heavy roofing and plaster ceiling finish. In assembling wood trusses, special care should be taken to achieve adequate nailing since the strength of trusses is, to a large extent, dependent on the fastness of the connection between members. Care should also be exercised in selecting materials for trusses. Lumber equal in stress value to No. 2 dimension shortleaf southern pine is suitable; any lower quality is not recommended. Trusses are one place where framing fasteners, such as Teco makes, can be very important.

ROOF VENTILATION

Adequate ventilation is necessary in preventing condensation in buildings. Condensation may occur in the walls, in the crawl space under the structure, in basements, on windows, etc. Condensation is most likely to occur in structures during cold weather when interior humidity is high. Proper ventilation under the roof allows moisture-laden air to escape during the winter heating season, and also allows the hot dry air to escape during the winter heating season. It also allows the hot dry air of the summer season to escape, which will keep the house cooler. The upper areas of a structure are usually ventilated by the use of louvers or ventilators. The various types of ventilators used are as follows:

1. Roof louvers (Fig. 48).

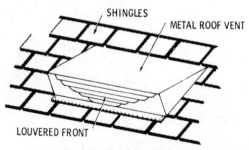

Fig. 48. Roof ventilator.

2. Cornice ventilators (Fig. 49).

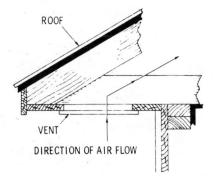

Fig. 49. Cornice ventilator.

3. Gable louvers (Fig. 50).

Fig. 50. Typical gable ventilator.

4. Flat-roof ventilators (Fig. 51).

Fig. 51. Typical flat roof vent.

169

5. Crawl-space ventilators (Fig. 52).

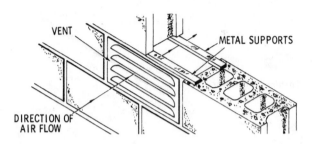

Fig. 52. Typical crawl space ventilator.

6. Ridge ventilators (Fig. 53).

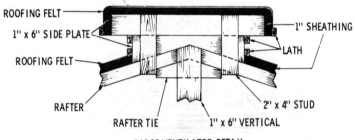

RIDGE VENTILATOR DETAIL

Fig. 53. Ridge-type ventilator.

The upper structure ventilator, one of the most common methods of ventilating, is by the use of wood or metal louver frames. There are many types, sizes, and shapes of louvers. The following points should be considered when building or installing the various ventilators.

1. The size and number of ventilators are determined by the size of the area to be ventilated.
2. The minimum net open area should be ¼ square inch for every square foot of ceiling area.
3. Most louver frames are usually 5 inches wide.
4. The back edge should be rabbeted out for a screen or door, or both.

5. Three-quarter-inch slots are used and spaced about 1¾ inches apart.
6. Sufficient slant or slope to the slots should be provided to prevent rain from driving in.
7. For best operation, upper structure louvers are placed as near the top of the gable roof as possible.

The crawl space installed in foundation structures should be well ventilated. Air circulation under the floor prevents excessive condensation, which causes warping, swelling, twisting and rotting of the lumber. These vents are usually called *foundation louvers;* some types are used for crawl spaces if ever needed. They are set in the foundation as it is being built. A good foundation vent should be equipped with a copper or bronze screen and adjustable shutters for opening and closing of the louvers. The size of the vents should be figured on the same basis as that used for upper structure ventilation.

SUMMARY

There are numerous forms of roofs and an endless variety of roof shapes. The frame of most roofs is made up of timbers called rafters. The terms used in connection with roofs are span, total rise, total run, unit of run, rise in inches, pitch, cut of roof, line length, and plumb and level lines.

Rafters are the supports for the roof covering and serve in the same manner as do joists for floors or do studs for the walls. Rafters are constructed from ordinary 2 × 6, 2 × 8, or 2 × 10 lumber, spaced 16 to 24 inches on center. Various kinds of rafters used in roof construction are common, hip, valley, jack, and octagon. The length of a rafter may be found in several ways— by calculation, with a steel framing square, or with the aid of a framing table.

Definite savings in material and labor through the use of pre-assembled wood roof trusses make truss framing an effective means of cost reduction in small buildings. In addition to cost savings, roof trusses offer advantages of increased flexibility for interior planning, and added speed and efficiency in site erection procedures.

Light wood trusses that permit substantial savings because they may be spaced 24 inches on center as compared to the usual 16-inch spacing have been developed. Aside from the direct benefits of reduced cost and savings in material and labor requirements, roof trusses offer special advantages in interior design, since the roof span can be increased without center supports. This offers unlimited advantages in floor areas for auditoriums and meeting rooms.

REVIEW QUESTIONS

1. Name the various kinds of rafters used in roof construction.
2. Name the various terms used in connection with roofs.
3. What is preassembled truss roofing? What are some advantages in this type of construction?
4. Why is adequate roof and crawl-space ventilation needed?
5. What are jack rafters and octagon rafters?

CHAPTER 11

Skylights

A skylight is any window placed in the roof of a building or ceiling of a room for the admission of light and/or ventilation. Skylights are essential to lighting and ventilating top floors where roofs are flat, such as in factories, and where there is not much side lighting. In the home, they are often placed at the top of stairs or in a room where no side window is available.

Fig. 1 shows a simple hinged skylight and detail of the hinge. The skylight may be operated from below by the control device, which has an adjustment eye in the support for securing the hinged sash at various degrees of opening. A skylight is often placed at the top of a flight of stairs leading to the roof, the projecting structure having framed in it the skylight and a doorway, as in Fig. 2.

Where fireproof construction is required, skylights are made of metal. Side-pivoted sash ones are shown in Fig. 3; this is a type desirable for engine and boiler rooms where a great amount of steam and heat is generated, and where a storm coming through them would do little harm if the skylights were thoughtlessly left open. The sash may be operated separately by pulleys or all on one adjuster. The ends may be stationary.

Wire glass should be used for factory skylights so that, if broken, it will not fall and possibly injure someone below. Wire glass is cast with the wire netting running through its center and is manufactured in many styles and sizes.

Skylights are becoming more popular in residential dwellings. They come double insulated (Fig. 4) and can be opened for ventilation. They are set in waterproof frames, which are perma-

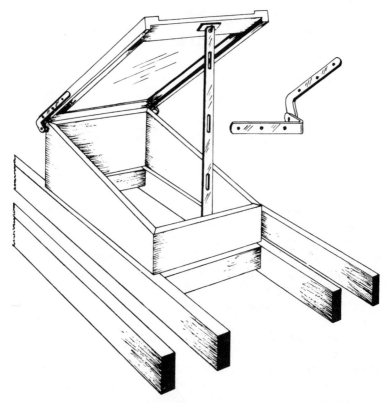

Fig. 1. Hinged skylight framed into roof.

nently installed and sealed in the roof. A skylight can often help heat a house; a lot of radiant heat from the sun gets through.

SUMMARY

Skylights are becoming more popular in residential dwellings. A skylight comes double insulated and can be opened for ventilation.

A skylight is often placed at the top of a flight of stairs leading to the roof or in an inside room where no side windows are available. Skylights may also be used in areas where the sun penetrates the glass for heating rooms. In some cases, the sun is used to heat water stored in ceiling tanks.

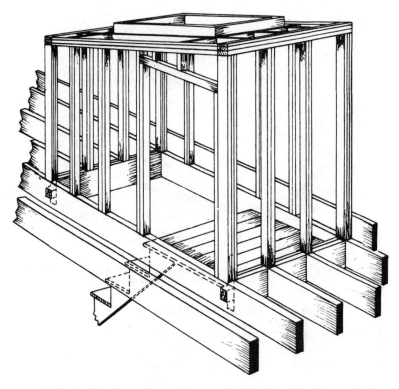

Fig. 2. View of entrance to roof, or projection framework containing framed opening for skylight and doorway.

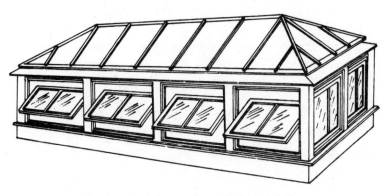

Fig. 3. Metal fireproof ventilating skylight.

175

Fig. 4. Skylight in a residence. It floods dark areas with light and can even help to heat the home.

REVIEW QUESTIONS

1. What are some of the advantages of a skylight?
2. Why are skylights used in some dwellings?
3. Where are skylights placed in dwellings?

CHAPTER 12

Porches and Patios

The terms, stoop, porch, and veranda are usually applied without distinction to mean a covered structure forming an outside entrance to a building.

TYPES OF CONSTRUCTION

A distinction should be made between the terms above and the word stoop, which means an uncovered platform at the door of a house, usually having steps with a hand rail at each side. A stoop is virtually a primitive porch without a cover.

Porches—Wooden porch floors are still built, and fundamental rules still obtain: Unless the wood is pressure treated with Penta or some other preservative, it will soon decay and need replacing. The porch may or may not have a roof, which will depend on the building design. This does not pose a problem because awnings are available.

Patios—The patio, and the whole concept of barbecue-backyard living, has taken the place of porch living. There are many different ways to create a patio area. You can build a slab, install flagstone or brick or, like many people, build a deck either on the ground level or elevated (Fig. 1). This, of course, will depend on the construction of the house. Some basic ideas follow in text and illustration.

Laying a Concrete Patio Floor—As always, the first stage is to decide on the boundaries of the area and to mark them out with

177

Fig. 1. Many people like a deck, such as this one between house and garage, for outdoor living. Courtesy of Western Wood Products Assn.

stakes and string. If the paved area is to be large, it should be divided into smaller sections and leveled, as shown in Fig. 2.

The stakes are driven into the ground until the tops are level in both directions. This is checked by a level placed on a straightedge across the stakes. The form boards are then nailed to and even with the tops of the outer stakes.

An interesting and attractive effect can be obtained by adding a design to your patio floor. Adding color to the concrete, and different finishing techniques to the floor are just a few ideas. And, as mentioned, brick or flagstone are two very popular materials often used for a patio floor.

Building a Small Patio Wall—Small patio walls provide an attractive informal appearance that can easily be maintained. There are a number of occasions when a low wall can provide the answer to the problem of privacy. Constructing such a wall is not a job that the inexperienced should be afraid of tackling. It is possible to purchase several types of precast block or brick that have an informal appearance suitable for a patio, as shown in Fig. 3.

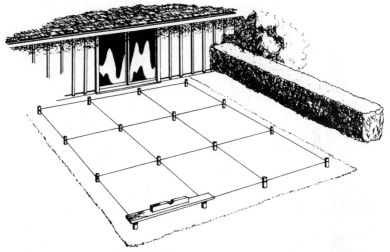

Fig. 2. Staking out and leveling a large area for concrete slab.

Depending on the circumstances, a wall can be single thickness, with or without a flat capping on top, or it can be a double-thickness wall joined at the ends, which can be filled with earth. These double walls should be at least three inches in thickness with a four-inch cavity. Up to the height of two feet, the wall will not need to be tied together. Alternatively, the wall can be single thickness with hollow piers at the ends forming a flower box or lamppost.

A tremendous variety of fencing materials is available, from a stockade fence, which gives complete privacy, to a ranch fence, which is just a series of timbers that define an area. Some of this material comes in handy, partially assembled sections; a trip to the lumberyard can be instructive.

Don't forget, in order for a patio to be used, there must be good access to it. There are many kinds of patio doors, wood and metal, that can be used (Fig. 4). These let in light, make rooms seem larger, and provide good patio access.

SUMMARY

Porches and patios make a house more livable. Today, the concept of patio living is a very popular one.

Fig. 3. A block patio wall.

In building a patio, first decide on the area and mark the boundaries with stakes and string. Many interesting and attractive effects can be obtained by adding a design to your patio floor. Adding color to the concrete or adding different finishes or designs are just a few ideas. Two very popular materials often used for patio or porch floors are brick or flagstone. Interesting and attractive designs can be obtained when using such material.

Many times a small patio wall or fence will provide an attractive informal appearance that can easily be maintained. The wall can be constructed with brick or concrete blocks in many attractive designs. In many cases, fencing can be installed. A

*Fig. 4. Patio door gives access to outdoor living, brings outdoors inside.
Courtesy of Andersen.*

fence is quick and easy to install and can add charm and beauty
to your patio.

REVIEW QUESTIONS

1. Why are wooden porch floors a thing of the past?
2. What type of patio wall is generally constructed?

CHAPTER 13

Chimneys, Fireplaces, and Stoves

As mentioned earlier, chimneys and fireplaces are not usually constructed by the carpenter, although many tradesmen are adept at more than one skill. But in many cases, carpentry is required for proper installations particularly when dealing with prefab units. To the general builder, contractor, architect, prospective homeowner, and do-it-yourselfer the construction features of these masonry units are of interest, and stoves in particular can be great energy savers.

CHIMNEYS

The proper construction of a chimney is more complicated than the average person realizes. Not only must it be mechanically strong, it must also be pleasing to the eye and correctly designed to provide a proper draft. A chimney can be too large as well as too small.

Conditions for Proper Draft

The power that produces a natural draft in a chimney is due to the small difference in weight between the hot gases in the flue and the colder air outside. The hot gases are lighter and will therefore rise, providing a natural draft in the chimney.

A frequent cause of poor draft is that the peak of the roof extends higher than the top of the chimney. This causes downdrafts as the wind sweeps across the roof and swirls down on top of the chimney. To prevent this condition, make sure the top of the chimney is at least 2 feet above the peak of the roof, as in Fig. 1A. The height above a flat roof should not be less than 3 feet, as in Fig. 1B.

Another cause of poor draft is a restriction of some sort inside the chimney that offers a resistance to the free flow of gases.

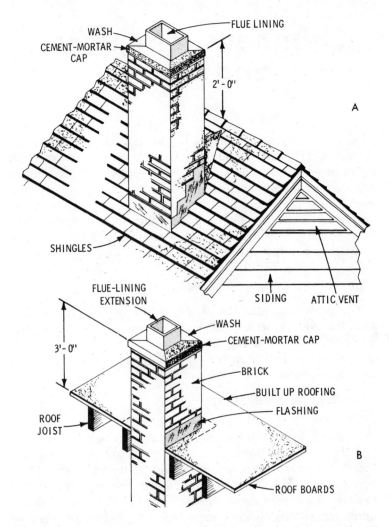

Fig. 1. Chimney heights: (A) above a gable roof (B) above a flat roof.

This restriction may be in the form of an offset that is too abrupt. Therefore, chimneys should be built with as little offset as possible. This not only provides a better draft but also prevents the accumulation of soot and ashes that may eventually block the chimney completely. Fig. 2 shows the effect of a gradual offset and the effect of an abrupt offset. Neither of these chimneys illustrates good design, however, because the offset should never be displaced so much that the center of gravity of the upper portion falls outside the area of the lower portion (Fig. 3). Otherwise, the unit will not be structurally sound and may develop cracks or, in extreme cases, fall.

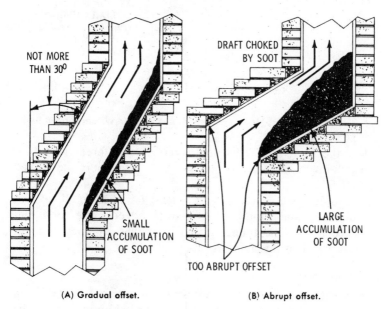

NOT MORE THAN 30°

DRAFT CHOKED BY SOOT

SMALL ACCUMULATION OF SOOT

LARGE ACCUMULATION OF SOOT

TOO ABRUPT OFFSET

(A) Gradual offset.

(B) Abrupt offset.

Fig. 2. An offset in a flue must not be too abrupt.

Foundations

When consideration is given to the amount of weight contained in even the smallest chimney, it becomes evident that a good foundation is necessary to support it properly. In no case should the foundation be less than 18 inches deep. If the chimney is built on an outside wall, the foundation must extend to

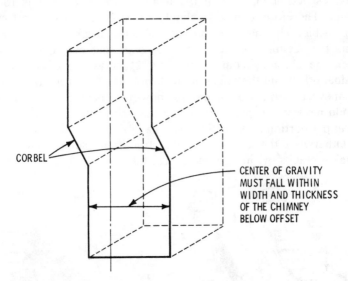

CORBEL

CENTER OF GRAVITY
MUST FALL WITHIN
WIDTH AND THICKNESS
OF THE CHIMNEY
BELOW OFFSET

Fig. 3. The offset (corbel) of a chimney must be kept within certain limits.

below the frost line. This depth varies in different localities, being as much as 4 feet in some northern sections of the United States.

The foundation should extend at least 6 inches on all sides of the chimney and should be of reinforced concrete for larger fireplace units. The foundations for smaller chimneys do not necessarily need this reinforcement.

Construction

A straight flue is the most efficient and easiest to build. If an offset is necessary, it should not be displaced by more than the amount shown in Fig. 3, and should form an angle of less than 30° with the vertical.

Common brick is the most widely used material in the construction of chimneys. When this type brick is laid in a single course, the wall of the unit will be 4 inches thick and should be lined with a fire-clay flue lining. *Never omit the lining or replace it with plaster.* The normal expansion and contraction of the walls will cause a plaster lining to crack, forming an opening to the outside of the chimney wall. This will result in a definite fire hazard. If a flue liner is not used, the wall should be no less than 8 inches thick, with the mortar joints carefully pointed. The plaster is applied to the inside of the flue as it is built. A bag filled with shavings may be placed in the flue and drawn up as the work progresses. If this bag fits the flue snugly, it will catch all the plaster droppings and will remove any mortar projecting into the opening. The use of a bag in this manner is also recommended when a flue liner is used, especially if there is an offset in the chimney.

The flue lining should extend the entire height of the chimney, projecting about 4 inches above the cap, as in Fig. 4. A slope should then be formed of cement within 2 inches of the top of the lining. This gives an upward direction to the wind currents across the top of the flue and tends to prevent rain and snow being blown in. For chimneys having more than one flue, at least a 4-inch course of brick must separate the liners. These liners come in various dimensions. Table 1 shows the sizes generally available. The length of each unit is normally 24 inches, although lengths from 20 to 30 inches are sometimes available.

FIREPLACES

Since the energy crunch, fireplaces have become popular. In fact, though, they normally do not lower fuel bills—they are not

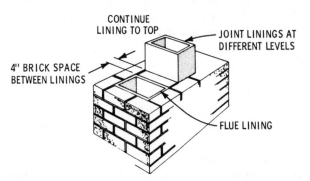

Fig. 4. Flue liners should extend to the top of the chimney and project about 4 inches. When two or more flues are present, they should be separated by 4'' of brick, and the joints of the liners should be staggered so they do not occur at the same level.

Table 1. Rectangular and Round Flue Linings

Rectangular flues		Round flues	
outside dimensions (inches)	net inside area (square inches)	inside diameter (inches)	net inside area (square inches)
4½ X 8½		6	28.3
4½ X 13		7	38.5
4½ X 18		8	50.3
8½ X 8½	52	9	63.6
8½ X 13	80	10	78.5
8½ X 18		12	113.
13 X 13	126	15	176.7
13 X 18	169	18	254.5
18 X 18	240	20	314.2
		24	452.4

efficient. If you want a fireplace, you should choose it more for the cheerful and cozy feeling that a crackling blaze in a fireplace gives. Fireplaces are considered appropriate in a living room, dining room, and bedroom.

There are a number of design features to observe to ensure that the fireplace performs satisfactorily. Fireplace size should match the room in which it is to be used, both from the standpoint of appearance and operation. If the fireplace is too small, it will not adequately heat the room—if too large, a fire that fills the combustion chamber will be much too hot for the room and will waste fuel.

The location of the chimney usually determines the location of the fireplace, and is too often governed by structural considerations only. A fireplace should be located to provide a reasonable amount of seclusion and should not, therefore, be near doors that carry traffic from one part of the house to another.

Fireplace Design

In designing a fireplace, the characteristics of the fuel to be burned should be considered. In addition, the design should harmonize with the room in both proportion and detail. In the days of the early settlers of our land, wood was plentiful and fireplaces with openings as large as 7 feet wide and 5 feet high were common. These were often used for cooking as well as heating. Needless to say, they consumed huge amounts of fuel and were more likely to be smoky than not.

Where cordwood (4 feet long) is readily available and practical (and promises to stay that way!), a 30-inch-wide opening is desirable for a fireplace. The wood is cut in half to provide pieces 24 inches long. If coal is to be burned, however, the opening can be narrower. Thirty inches is a practical height for any fireplace having an opening less than 6 feet wide. It will be found that, in general, the higher the opening, the greater is the chance the fireplace will smoke.

Another dimension to consider is the depth of the combustion chamber. Again, in general, the wider the opening, the deeper the chamber should be. A shallow opening will provide more heat than a deep one of the same width, but will accommodate a smaller amount of fuel. Thus, a choice must be made as to which is the most important—a greater depth, which permits the use of larger logs that burn longer, or a shallow depth, which takes smaller pieces of wood but provides more heat.

In small fireplaces, a depth of 12 inches will provide adequate draft if the throat is constructed properly, but a minimum depth of 16 to 18 inches is advisable to lessen the danger of burning fuel falling out on the floor. Screens of suitable design should be placed in front of all fireplaces.

A fireplace 30 to 36 inches wide is usually suitable for a room having 300 square feet of floor area. The width should be increased for larger rooms, with all other dimensions being changed to correspond to the sizes listed in Table 2. Following these dimensions will result in a fireplace that operates correctly.

Modified Fireplaces

Metal fireplaces are available in many different sizes and are designed to be set in place and concealed by the brick or stone used to lay up the fireplace. Fig. 5 shows one of these units. With this type of fireplace, the warm air is circulated throughout the room by the action of the air inlets and outlets provided in the metal unit. Cool air is drawn into the inlets at the bottom of the unit and is heated by contact with the metal sides. This heated air rises and is discharged through the outlets at the top. The inlets and outlets are connected to registers that may be located at the front or ends of the fireplace, or even in another room.

Table 2. Recommended Dimension for Fireplaces

Opening		Depth (inches)	Minimum back wall (horizontal) (inches)	Vertical back wall (inches)	Inclined back wall (inches)	Outside dimensions of standard rectangular flue lining (inches)	Inside diameter of standard round flue lining (inches)
Width (inches)	Height (inches)						
24	24	16-18	14	14	16	8½ X 8½	10
28	24	16-18	14	14	16	8½ X 8½	10
24	28	16-18	14	14	20	8½ X 8½	10
30	28	16-18	16	14	20	8½ X 13	10
36	28	16-18	22	14	20	8½ X 13	12
42	28	16-18	28	14	20	8½ X 18	12
36	32	18-20	20	14	24	8½ X 18	12
42	32	18-20	26	14	24	13 X 13	12
48	32	18-20	32	14	24	13 X 13	15
42	36	18-20	26	14	28	13 X 13	15
48	36	18-20	32	14	28	13 X 18	15
54	36	18-20	38	14	28	13 X 18	15
60	36	18-20	44	14	28	13 X 18	15
42	40	20-22	24	17	29	13 X 13	15
48	40	20-22	30	17	29	13 X 18	15
54	40	20-22	36	17	29	13 X 18	15
60	40	20-22	42	17	29	18 X 18	18
66	40	20-22	48	17	29	18 X 18	18
72	40	22-28	51	17	29	18 X 18	18

One advantage of this type of modified fireplace is that the firebox is correctly designed and proportioned, and has the throat damper, smoke shelf, and chamber already fabricated. All this provides a foolproof form for the masonry and greatly reduces the risk of failure and practically assures a smokeless fireplace. It must be remembered, however, that even though the fireplace is built correctly, it will not operate properly if the chimney is inadequate.

The quantity and temperature of the heated air discharged from the registers in this type of fireplace will circulate heat into the coldest corners of a room and will deliver heated air through ducts to adjoining or overhead rooms. For example, heat could be diverted to a bathroom from a fireplace in the living room. The amount of heat delivered by some of the medium-size units is equivalent to that from 40 square feet of cast-iron radiators in a hot-water heating system. This is sufficient to heat a 15' × 18'

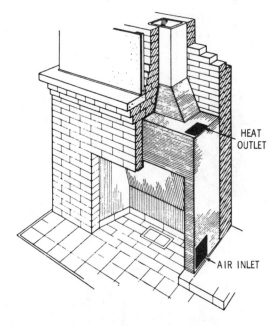

Fig. 5. A metal fireplace unit practically assures a smokeless fireplace.

room to 70°F when the outside temperature is 40°F. Additional heat can be obtained with some models by the use of circulation fans placed in the ductwork.

Fig. 6 shows another type of modified fireplace in which the circulated air is drawn in from outside through an air inlet *a*. The air is then heated as it rises through the back heating chamber *c* and tubes *t*, being discharged through register *b*. Combustion air is drawn into the fireplace opening *d*, and passes between the tubes and up the flue. Dampers are provided to close the air inlet and regulate the amount of combustion air allowed to pass into the flue. Fig. 7 shows a cross section of this type of modified fireplace.

Prefabricated Fireplaces

Prefabricated fireplaces are available in many designs and price ranges. These units fill the need for a low-cost fireplace that eliminates expensive masonry construction. This type of

191

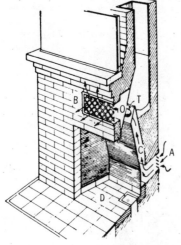

Fig. 6. A type of modified fireplace in which the air to be heated is drawn in from outside.

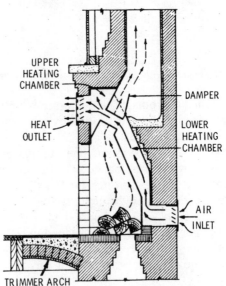

Fig. 7. A cross-sectional view of the modified fireplace shown in Fig. 6.

fireplace can be free-standing, mounted flush with the wall, or recessed into it, depending on the style selected. The unit shown in Fig. 8 has a cantilever designed hearth 15 inches above floor level and can burn wood up to 27 inches long. The outer shell is steel with stainless steel trim, while the firebox is formed of high-impact ceramic material. Fig. 9 shows a prefabricated fireplace that hangs from the ceiling.

The complete unit (except for the trim) is prime coated and ready to be painted with any interior paint to harmonize with the decorating scheme of the room. A flexible hearth screen is provided to prevent sparks from flying out of the firebox into the room.

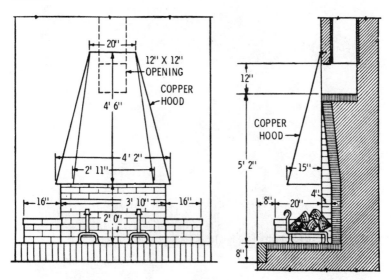

Fig. 8. A prefabricated fireplace that requires a minimum of masonry.

Fireplace Construction

Fireplace construction is an exacting art. Preformed metal units may be used, or the complete fireplace can be built of masonry materials. Fig. 10 shows the details of a typical fireplace. This, along with the following essentials, should prove helpful in the design and construction of any type of fireplace.

1. The flue must have sufficient cross-sectional area.
2. The throat must be constructed correctly and have a suitable damper.

Fig. 9. A free-standing fireplace can be located anywhere you wish.

3. The chimney must be high enough to provide a good draft.
4. The shape of the firebox must be such as to direct a maximum amount of radiated heat into the room.
5. A properly constructed smoke chamber must be provided.

Dimensions—Table 2 lists the recommended dimensions for fireplaces of various widths and heights. If a damper is installed, the width of the opening will depend on the width of the damper frame and slope of the back wall.

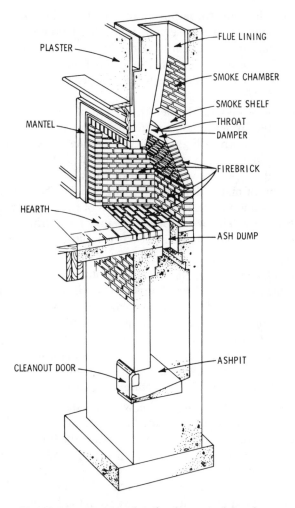

Fig. 10. Construction details of a typical fireplace.

The width of the throat is determined by the opening of the hinged damper cover. The full damper opening should never be less than the cross-sectional area of the flue. When no damper is used, the throat opening should be 4 inches for a fireplace not exceeding 4 feet in height.

Hearths—The hearth in conventional fireplaces should be

195

flush with the floor to permit sweepings to be brushed into the fireplace opening. If a basement is located below the fireplace, an ash dump on the hearth near the back of the firebox is convenient. This dump consists of a metal frame about $5'' \times 8''$ with a pivoted plate through which the ashes can be dropped into a pit below.

With wooden floors, the hearth in front of the fireplace should be supported by masonry trimmer arches or other fire-resistant material, as shown in Figs. 7 and 10. The hearth should project at least 16 inches from the fireplace breast and should be constructed of brick, stone, terra cotta, reinforced concrete, or other fireproof material at least 4 inches thick. The length of the hearth should not be less than the width of the fireplace opening plus 16 inches. The wood centering under the trimmer arches may be removed after the mortar has set, although it may be left in place if desired.

Wall Thickness—Fireplace walls should never be less than 8 inches thick if of brick, or 12 inches thick if of stone. The back and sides of the firebox may be laid up with stone or hard-burned brick, although fire brick in fire clay is preferred. When firebricks are used, they should be laid flat with their long edges exposed to lessen the danger of their falling out. If they are placed on edge, however, metal ties should be built into the main brickwork to hold the firebrick in place.

Jambs—The jambs should be wide enough to give stability and a pleasing appearance to the fireplace, and are usually faced with ornamental brick, stone, or tile. When the fireplace opening is less than 3 feet wide, 12- or 16-inch jambs are usually sufficient, depending on whether a wood mantel is used or if the jambs are of exposed masonry. The edges of a wood mantel should be kept at least 8 inches away from the fireplace opening. Similar proportions of the jamb widths should be kept for wider openings and larger rooms.

Lintels—Lintels are used to support the masonry over the fireplace openings. These lintels may be $\frac{1}{2}'' \times 3''$ bars or $3\frac{1}{2}'' \times 3\frac{1}{2}'' \times \frac{1}{4}''$ angle iron. Heavier lintels are required for wider openings.

If a masonry arch is used over the opening, the jambs should be heavy enough to resist the thrust imposed on them by the arch. Arches used over openings less than 4 feet wide seldom sag, but wider openings will cause trouble in this respect unless special construction techniques are employed.

Throats—The sides of the fireplace should be vertical up to the throat or damper opening, as shown in Fig. 11. The throat should be 6 or 8 inches or more above the bottom of the lintel and have an area that is not less than that of the flue. The width of the throat should be the same as the width of the fireplace opening. Starting 5 inches above the throat, the sides should be drawn in to equal the flue area.

Proper throat construction is necessary if the fireplace is to perform correctly, and the builder must make certain that the side walls are laid vertically until the throat has been passed and that the full length of opening is provided.

Smoke Shelf and Chamber—The smoke shelf is made by set-

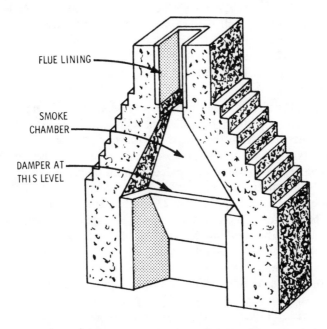

FLUE LINING

SMOKE CHAMBER

DAMPER AT THIS LEVEL

Fig. 11. View of the fireplace and flue opening.

ting the brickwork back at the top of throat to the line of the flue wall for the full length of the throat. The depth of the shelf may vary from 6 inches to 12 inches or more, depending on the depth of the fireplace.

The smoke chamber is the space extending from the top of the throat up to the bottom of the flue proper, and between the side walls. The walls should be drawn inward at a 30° angle to the vertical after the top of the throat is passed. These walls should be smoothly plastered with cement mortar not less than ½ inch thick.

Damper—A properly designed damper provides a means of regulating the draft and prevents excessive loss of heat from the room when the fireplace is not being used. A damper consists of a cast-iron frame with lid hinged so that the width of the throat opening may be varied from a completely closed to a wide-open position.

Regulating the opening with the damper prevents waste of heat up the chimney. Closing the damper in the summer keeps flies, mosquitos, and other insects from entering the house through the chimney. In houses heated by furnaces or other heating devices, lack of a fireplace damper may interfere with uniform heating of the house, whether or not there is a fire in the fireplace.

Flue—The area of lined flues should be at least $^1/_{12}$ the area of the fireplace opening, provided the chimney is at least 22 feet high, measured from the hearth. If the flue is shorter, or is unlined, its area should be at least $^1/_{10}$ the area of the fireplace opening. For example, a fireplace with an opening of 10 square feet (1440 square inches) needs a flue area of approximately 120 square inches if the chimney is 22 feet tall or 144 square inches if less than 22 feet. The nearest commercial size of rectangular lining that would be suitable is the 13″ × 13″ size. A 15-inch round liner would also be suitable.

It is seldom possible to obtain a liner having the exact cross-sectional area required, but the inside area should never be less than that required. Always select the next larger size liner in such a case. Tables 1 or 3 can be used to select the proper size liner or for determining the size of fireplace opening for an existing flue. The area of the fireplace opening in square inches is obtained by multiplying the width by the height, both in inches.

Stoves

Today, because of the energy crunch, many people are buying stoves. These are far more efficient than fireplaces simply because they are not open to the air. They do not use up a great deal of oxygen, which is warmed air, to keep going, as a fireplace does.

Stoves may be bought for burning coal as well as wood. They come in a variety of styles, with the airtight stove (Fig. 12) generally conceded to be the most fuel efficient. As with the fireplace, proper installation of a stove is a must. Manufacturers provide installation instructions, but it is also wise (and usually legally necessary) to consult with a local building code authority on the construction of a stove.

SUMMARY

Proper construction of a chimney is more complicated than the average person realizes. A chimney can be too large as well as too small. Not only must it be mechanically strong, it must

Table 3.   Sizes of Flue Linings for Various Size Fireplaces
(Based on a flue area equal to one-twelfth the fireplace opening)

Area of fireplace opening (square inches)	Outside dimensions of standard rectangular flue lining (inches)	Inside diameter of standard round flue lining (inches)
600	8½ X 8½	10
800	8½ X 13	10
1,000	8½ X 18	12
1,200	8½ X 18	12
1,400	13 X 13	12
1,600	13 X 13	15
1,800	13 X 18	15
2,000	13 X 18	15
2,200	13 X 18	15
2,400	18 X 18	18
2,600	18 X 18	18
2,800	18 X 18	18
3,000	18 X 18	18

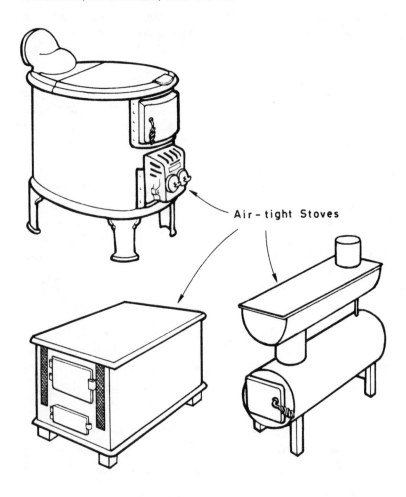

Air-tight Stoves

Fig. 12. The most heat efficient type of stove is the air-tight.

also be properly designed to provide the proper draft. A frequent cause of poor draft is that the peak of the roof extends higher than the top of the chimney.

A good foundation is necessary to support a properly built chimney. If a chimney is built on an outside wall, the foundation must extend well below the frost line. If a large fireplace is built in connection with the chimney, the foundation should extend at least 6 inches on all sides with reinforced concrete.

A straight flue is the most efficient design constructed. If an offset is necessary, it should not form an angle of more than 30 degrees. The most widely used material in chimney construction is brick or stone, and should never be designed without the use of a flue liner. The flue liner should always extend the entire height of the chimney and project about 4 inches above the chimney cap.

All fireplaces should be built according to all rules and regulations to perform satisfactorily. Size should always match room area from the standpoint of appearance and operation.

For the person interested in energy efficiency, there are stoves. The airtight design is best.

REVIEW QUESTIONS

1. Why should chimneys always be built in a straight line?
2. How high should a chimney be built above the roof? Why?
3. Explain the purpose of the flue lining.
4. What should be considered when designing a fireplace?
5. Explain the purpose of the hearth, throat, and damper.

CHAPTER 14

Insulation and Energy Saving

In the past few years, ever since the energy crunch began, there has been a tremendous interest in energy saving, both on new work and existing structures. Carpenters, contractors, builders, and do-it-yourselfers have found that, over time, using certain materials and techniques can result in big savings.

What follows is a roundup of energy savers for new and existing buildings. In most cases the suggestions can be readily adopted. But it is always a good idea to have discussions with knowledgeable people, such as experts at local utility companies, to see what works best. For example, 12-inch-thick insulation may be just the thing for frigid New Hampshire winters, but it would hardly be practical, in terms of original cost, for insulating a stucco home south of Los Angeles.

INSULATION — R FACTORS

A key concept to remember when selecting insulation is the R factor of the material. *R* stands for resistance: the ability of the material (technically the stratified air spaces) to resist or retard heat. In the winter this resistance works by keeping warm air inside the home. In the summer it keeps hot air out.

The critical time is winter, when the expensive fuel used to heat that warm air may be traveling out of the house by conduction, making the furnace work harder—and using up more fuel.

Cold air infiltration is also an energy factor. If the house has cracks and openings where cold air can sneak in (and warm air escape), this will also result in higher fuel costs. Such openings must be plugged up.

Insulation comes in a variety of types, generally ranging from about R-4 to R-38. (All insulation has its R factor stamped on it.) The R-4 material could be urethane boards applied to basement walls, the R-38 12-inch-thick fiberglass batts. The following are the materials available:

Rigid Board Insulation—This is available in various sizes in board form. It may be urethane, fiberglass, Styrofoam, or composition board (Figs. 1 and 2). It is easily worked with standard hand tools, and some contractors favor it as sheathing. While the R values are not high compared with other kinds of insulation, they are better than those of board or plywood. Rigid board insulation is also handy to use in the basement, where it can be glued directly to walls. Before using it, however, you should

Fig. 1. A block or masonry wall is a good candidate for Styrofoam or urethane board insulation. Courtesy of Dow.

Fig. 2. *Material is glued on. Gypsum board must be used over plastic if material is combustible. Courtesy of Certain-Teed*

check for fire restrictions. Some rigid board insulation is flammable and must be covered by a nonflammable material before any other flammable material, such as wood paneling, can be installed.

Siding with Insulation—Some siding, chiefly aluminum, has built-in insulation. The R factors vary, however, and should be checked out before any material is purchased.

Batts and Blankets—Batts are flexible insulation sections cut to fit between standard studs (on 16-inch centers) that are about 8 feet long. Blankets are the same width as batts but come in virtually infinite lengths (up to 100 feet). Batts and blankets may be faced on both sides with a paperlike material, and one side will have a vapor barrier—either aluminum or kraft paper—or may be unfaced (Fig. 3). Mineral fiber is the filler for such insulation: fiberglass, such as that made by Owens Corning, or mineral wool. On the edges of batts and blankets are strips, which are stapled to framing members.

Batts and blankets are favored by contractors and do-it-yourselfers alike. This insulation is used on exposed stud walls, on attic floors (between joists), and on open ceilings. It ranges in R factor from around R-4 to R-38.

Fill-Type Insulation—Another insulation is the fill type (Fig. 4), which comes as a granular material (Vermiculite) or a macerated paper (cellulose). This material may be poured in place, such as between attic joists, but its primary use is to be blown into areas inaccessible for installing other insulation, mainly walls in finished homes. Sections of siding are removed, holes are drilled between pairs of studs, a hose is inserted, and the material is blown into place. It fills up the cavities and insulates.

Foam—Like fill-type insulation, foam is designed to be used on finished work that is inaccessible to the batts or blankets. The foam is mixed and is applied as a liquid. It foams up, filling the spaces, then becomes hard. There has been controversy about foam insulation. Complaints have been voiced—and indeed the government has investigated them—about medical problems caused by the fumes of foam, and its use has been banned in many places. In fairness to foam makers and installers though, a wise precaution is to know the type of foam an installer plans to use. Some foams are considered harmless but may suffer from a blanket indictment of all foam. Foam is practical, too, because nothing has a higher R value, inch for inch, than foam.

REDUCING LEAKAGE AROUND DOORS AND WINDOWS

A large amount of the heat delivered to rooms is lost by air leakage around doors and windows. Another factor in heat loss is heat leakage through window or door glass or exterior doors.

These various heat losses may be greatly prevented by:

1. Calking.

Fig. 3. Blanket insulation. It goes up to R-38, 12 inches thick. Courtesy of Certain-Teed.

Fig. 4. Fill insulation (mineral wool) being blown into attic. Courtesy of Mineral Wool Mfg. Assn.

207

2. Weather strips.
3. Storm or double-glass windows.
4. Storm doors.

Though the house may have been tight when built, a few years of usage will sometimes permit a considerable amount of leakage. In addition, small cracks often develop in masonry walls which, in time, may become sufficiently large to permit a considerable air leakage.

Calking

The most common remedy for air leakage is found in calking or plugging the leaks. A variety of calks are available, from oil base to butyl rubber. Calk is easiest to use in cartridges, which are slipped into a calking gun (Fig. 5). The gun extrudes a thin bead of calk, which will effectively seal off cracks. Generally, cracks form where dissimilar materials meet, such as on wood trim and siding. For deep cracks, oakum or mineral fiber can be packed into the crack before the calk is used.

Weatherstripping

Weatherstripping is another excellent idea. Weatherstrip comes in a variety of types and is designed to be used where there is a natural opening in the house, such as beneath the garage or exterior doors, or where windows meet framework. You can use felt material, which can be tacked in place to V-shaped strips that have one adhesive edge. Also available is a moldable weatherstrip and adhesive-backed felt. Whichever kind you use, weatherstripping is an energy saver.

Window and Door Glazing

A large amount of heat can be lost through window glass, and so storm windows and storm doors are an excellent idea. Indeed, they are just as important as insulation. If you find that you cannot afford storm windows this winter, you can use polyethylene sheeting and make your own. Polyethylene is as effective as glass, though it is only temporary, of course.

An even better solution for windows is double- and even triple-glazed windows. A dead air space is trapped between the

Fig. 5. Calk is important in the fight to save energy. Courtesy of United Gilsonite.

layers of glass and retards heat transmission. Such glazing is initially expensive, but it pays for itself over time.

WHERE TO INSULATE

A basic question, of course, is where to insulate. Generally, all walls, ceilings, roofs, and floors that separate heated from unheated spaces should be insulated. Insulation should be placed on all outside walls and in the ceiling as shown in Fig. 6. In houses with unheated crawl spaces, insulation should be placed between the floor joists. If a blanket type of insulation is used, it should be well supported with galvanized mesh wire or by a rigid board. The most important area to insulate, though, is

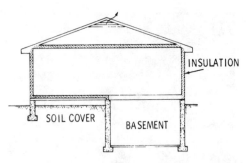

Fig. 6. Areas to be insulated. Attic or top-floor ceiling is most important.

the attic, simply because heat rises. The greatest amount of heat is lost through the attic.

In 1½-story houses, insulation should be placed along all areas that are adjacent to unheated areas, as shown in Fig. 7. These include stairways, dwarf walls, and dormers. Provisions should be made for ventilation of the unheated areas. Where attic storage space is unheated and a stairway is included, insulation should be used around the stairway as well as in the first floor ceiling, as shown in Fig. 8. The door leading to the attic should be weatherstripped to prevent the loss of heat. Walls adjoining an unheated garage or porch should be well insulated.

In houses with flat or low-pitched roofs, shown in Fig. 9, insulation should be used in the ceiling area with sufficient space allowed above for a clear ventilating area between the joists. Insulation should be used along the perimeter of a house built on

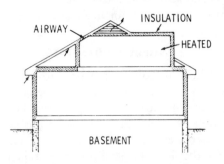

Fig. 7. Method of insulating 1½-story house. Basement wall could also be insulated with rigid board.

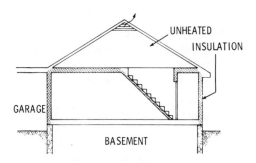

Fig. 8. *Illustrating a method of shielding a heated area from an unheated area by insulating.*

a slab. A vapor barrier should be included under the slab. Insulation, as mentioned, can be used to improve comfort within the house during hot weather. Surfaces exposed to the direct rays of the sun may attain temperatures of 50°F or more above shade temperature and, of course, tend to transfer this heat toward the inside of the house. Insulation in roofs and walls retards the flow of heat. And consequently, less heat is transmitted through such surfaces.

Where any cooling system is used, insulation should be used in all exposed ceilings and walls in the same manner as for preventing heat loss in cold weather. Of course, where air conditioning is used, the windows and doors should be kept closed during periods when outdoor temperatures are above inside temperatures. Windows exposed to the sun should be shaded with awnings.

Ventilation of attic and roof space is an important adjunct to insulation. Without ventilation, an attic space may become very hot and hold the heat for many hours. Obviously, more heat will

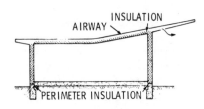

Fig. 9. *Insulation of a typical flat roof house.*

be transmitted through the ceiling when the attic temperature is 150°F than when it is 120°F. Ventilation methods suggested for protection against cold-weather condensation apply equally well to protection against excessive hot-weather roof temperature. (A detailed discussion of ventilation is given later.)

INSTALLATION OF INSULATION

Blanket or batt insulation should be installed between framing members so that the tabs lap all the face of the framing members, or on the sides. The vapor barrier should face the inside of the house. Where there is no vapor barrier on the insulation, a vapor barrier should be used over these unprotected areas (Fig. 10). A hand stapling machine is usually used to fasten insulation tabs and vapor barriers in place. A vapor barrier should be used on the warm side (the bottom, in case of ceiling joists) before insulation is placed. Figs. 11 through 17 show a number of other installation tips.

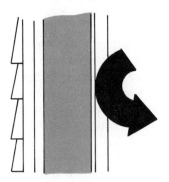

Fig. 10. The right way to install a vapor barrier.

VAPOR BARRIERS

Most building materials are permeable to water vapor. In cold climates during cold weather such vapor, generated in the house from cooking, dishwashing, laundering, bathing, humidifiers, and other sources, may pass through wall and ceiling materials and condense in the wall or attic space, where it may subsequently do damage to exterior paint and interior finish, or may

Fig. 11. Insulation in attic starts with pressing it in place under eaves; then, roll toward center of area. Courtesy of Certain-Teed.

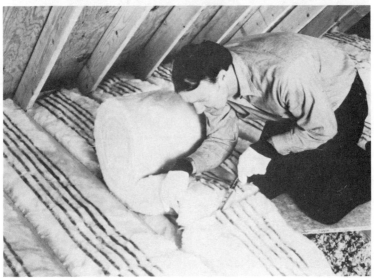

Fig. 12. Insulation can be cut with sharp knife. Courtesy of Certain-Teed.

213

even cause decay in structural members. As a protection, a material highly resistive to vapor transmission, called a *vapor barrier,* should be used on the warm side of any unfaced insulation, as mentioned earlier. The most widely used vapor barrier material is polyethylene sheeting. This is sold by the foot off rolls and can be attached easily.

Some batt and blanket insulations have a vapor barrier on one side of the blanket. Such material should be attached on the sides or faces of the studs. Where polyethylene is used, it should be applied vertically over the face of the studs and stapled on. The wall material is then applied over the vapor barrier. Vapor barriers should be cut to fit tightly around electric outlets and switch boxes. Any cold-air returns to the furnace in outside walls should be lined with tin.

VENTILATION

It is an excellent idea to button up a house with insulation and other energy-saving steps. But it is also important to ventilate a

Fig. 13. Insulation may be supplemented by installing additional blankets on existing material.

Fig. 14. *Above sill and between joists is one area many people forget to insulate. Tuck insulation in place.*

house properly to avoid condensation and high temperatures.

The attic and crawl spaces, in particular, deserve attention. Vents—power types and plain standard vents—let moisture escape, and in summer the moving air in an attic tends to keep it cooler. Indeed, an attic can reach an almost unbelievable 140°F in summer.

Attics

At least two vent openings should always be provided. They should be located so that air can flow in one vent and out the other. A combination of vents at the eaves and gable vents is better than gable vents alone. A combination of eave vents and continuous ridge venting is best (Fig. 18).

Fig. 15. Some manufacturers, such as Certain-Teed, recommend that new blankets of insulation be laid across existing material. If new material has a vapor barrier it should be slashed—opened up.

An attic should have a minimum amount of vent area, as follows:

Combinations of eave vents and gable vents without a vapor barrier—1 square foot inlet and 1 square foot outlet for each 600 square feet of ceiling area with at least half the vent area at the tops of the gables and the balance at the eaves.

Gable vents only without a vapor barrier—1 square foot inlet and 1 square foot outlet for each 300 square feet of ceiling area.

Gable vents only with a vapor barrier—1 square foot inlet and 1 square foot outlet for each 600 square feet of ceiling area.

Crawl Spaces

For a crawl space that is unheated, there should be at least two vents opposite one another. The minimum opening size, with moisture seal (4-mil or thicker polyethylene sheeting or

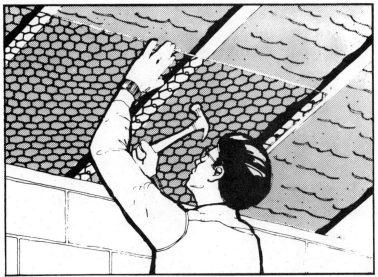

Fig. 16. Floors above unheated spaces should be insulated. Wire netting can be used to retain material.

55-lb. asphalt roll roofing lapped 3 inches) on the ground should be 1 square foot for every 1500 feet of crawl space.

If there is no moisture seal, then there should be an opening of 1 square foot for every 150 square feet.

Covered Spaces

When attic or crawl space vents are covered by screening or other materials, this results in a reduction in the size of the opening. Openings sizes should be increased, as indicated by the following:

Covering	*Size of Opening*
$1/4''$ hardware cloth	$1 \times$ net vent area
$1/4''$ hardware cloth and rain louvers	$2 \times$ net vent area
$1/8''$ screen	$1^1/4 \times$ net vent area
$1/8''$ screen and rain louvers	$2^1/4 \times$ net vent area
$1/16''$ screen	$2 \times$ net vent area
$1/16''$ screen and rain louvers	$3 \times$ net vent area

Fig. 17. Product from Du Pont called Tyvek is being used to cut air infiltration in an attic. Product is stapled into place.

SUMMARY

Insulation is a material used to prevent the high outside temperature from entering the house during the summer and the inside higher temperature from leaving the building during the winter.

The primary reasons for the application of insulating materials in homes are economy and comfort. When insulation is properly installed, it keeps the interior of the house comfortable with a minimum of fuel consumption. Also, in the summer months, insulation helps keep rooms in all parts of the house comparatively

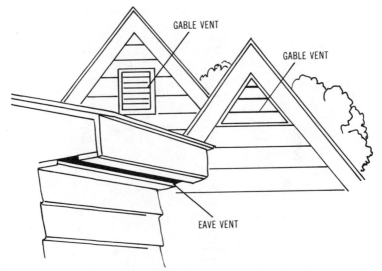

GABLE VENT

GABLE VENT

EAVE VENT

Fig. 18. Vent arrangements.

cool and comfortable. If air conditioning is used during the hot season, the system may be operated with more economy if the building is properly insulated.

There are several insulating materials. Among the more common materials used for insulation in homes are batts and blankets.

Fill-type insulation materials are powdered, granulated, or shredded, and come in bulk lots. This type of insulation can be applied in houses under construction or in those already completed, by pouring or blowing it into the space between the framing members. Fill insulation is generally placed between studs, ceiling joists, and attic joists.

The best insulation and its application will be of little help if air leakage around doors and windows is not controlled. The most common remedy against this air leakage is found in calking or filling the leaks. Many calking compounds are manufactured for various building materials.

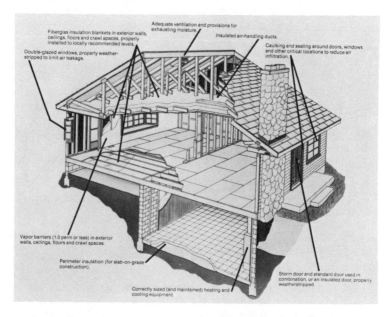

Fiberglas insulation blankets in exterior walls, ceilings, floors and crawl spaces, properly installed to locally recommended levels.

Double-glazed windows, properly weather-stripped to limit air leakage.

Adequate ventilation and provisions for exhausting moisture.

Insulated air-handling ducts.

Caulking and sealing around doors, windows and other critical locations to reduce air infiltration.

Vapor barriers (1.0 perm or less) in exterior walls, ceilings, floors and crawl spaces.

Perimeter insulation (for slab-on-grade construction).

Storm door and standard door used in combination, or an insulated door, properly weatherstripped.

Correctly sized (and maintained) heating and cooling equipment.

Fig. 19. An energy-conscious home. Courtesy of Owens-Corning.

REVIEW QUESTIONS

1. Name the various insulating materials.
2. How is fill insulation applied in a house?
3. What is a vapor barrier?
4. What is the purpose of insulation?
5. What is used to control air leaks?

CHAPTER 15

Solar Energy

One area that is increasingly likely to require the services of carpenters and builders, and an area in which more and more do-it-yourselfers are getting involved, is solar energy. Thus it is important for the craftsman to have a fundamental understanding of the systems involved.

The basic heating plant in any solar system is, of course, the sun. But one form of the system is active—hardware is required, as we will detail later—while the other is passive, and no moving hardware is needed.

Fig. 1. A solar house can look good, as this one shows. Courtesy of Acorn
Structures.

PASSIVE SYSTEMS

Passive-system solar buildings absorb and store solar heat in masonry floors, walls, or room dividers, or in water-filled tubes, drums, or other containers. Enough heat can be stored by a well-designed passive building to keep it warm after the sun has set and through the night, and possibly for a day or two of sunless weather.

The location of the storage mass in a passive building depends upon which passive method is used. In the popular "direct gain" approach (Fig. 2), the room collects sunlight through large south-facing windows. Solar radiation strikes the darkened surface of the thick masonry walls and floors, where it is absorbed as heat. The heat remains in the masonry until the air in the room cools. The heat then radiates into the room, drawn, as heat always is, from a warmer to a cooler spot until both places are of equal temperature. The surface area of storage exposed to direct sunlight in a "direct gain" building should be at least three times the area of the south-facing window. Storage walls and floors should be between 4 and 8 inches thick. An adequately sized storage mass prevents extensive overheating during the day and provides warmth through the night. It is important to remember that storage floors cannot be carpeted and that walls cannot be covered. These surfaces must be exposed if they are to absorb and radiate heat effectively.

In Fig. 3 the storage is called a Trombe wall (originally Trombe-Michel, after the two French scientists who developed it). The outside surface of the wall is painted black to absorb heat. Gradually migrating through the wall, the heat reaches the room later in the day and continues to radiate into the room through the night. For a Trombe wall, the area of the window and the area of the storage wall's surface should be the same. The wall should be from 8 to 16 inches thick.

Most Trombe walls also have vents at the floor and ceiling. While the sun shines, the heated air in the space between the glazing and the wall rises and enters the room through the upper vents, drawing cooler air into the space through the lower vents. This natural circulation of heat continues as long as the sun shines.

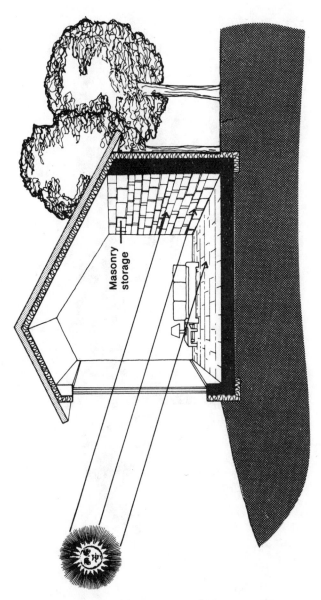

Fig. 2. "Direct gain" passive solar energy system.

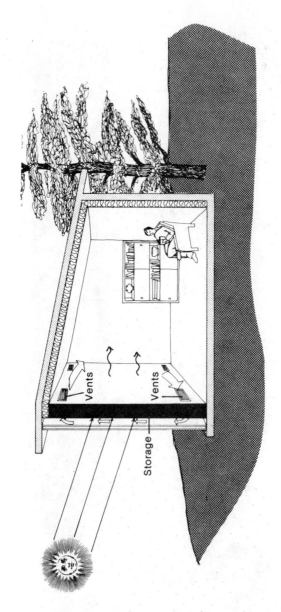

Fig. 3. Trombe wall, another passive system.

Water-filled containers (fiberglass tubes, 55-gallon drums, etc.) can be used to create a "water wall"; water has about twice the heat storage capacity of masonry of equivalent volume.

A variation on the water wall is the roof pond (Fig. 4). Roof ponds are essentially waterbeds made of sturdy plastic. They rest on a special ceiling structure and are covered and insulated on winter nights by a movable roof. During the day the mechanically operated roof is opened, exposing the roof pond to the sun so that it can absorb heat and later radiate it to the living area.

When an attached greenhouse (Fig. 5) is used to collect solar heat, the storage may include the masonry floor of the greenhouse, the wall separating the greenhouse from the main building, and water containers, such as the 55-gallon drums shown in the illustration. The wall may be either masonry or water-filled containers, and it may include windows for heating with direct sunlight.

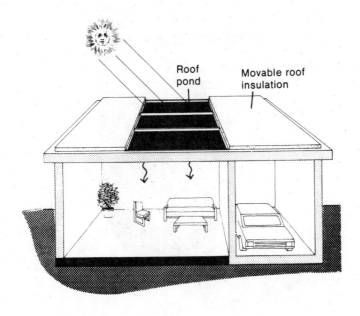

Fig. 4. A roof pond. It uses water to heat.

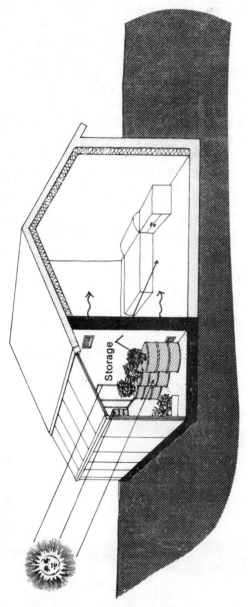

Fig. 5. Greenhouse is used to collect heat.

ACTIVE SYSTEMS

In active systems, air, water, or other liquid passes through the collector, which is warmed by the sun's rays. It then is routed by tubing of some sort to a storage tank or bin (in the case of air) where it is stored until use.

The components of an active system are generally housed in the basement or in a special utility area. Active-system storage is linked to both rooftop flat-plate collectors (Fig. 6) and a mechanical heat distribution system that makes use of pumps or air-handling units.

Ideally, storage bins or tanks for active systems should be located next to or under those rooms that require the greatest amount of heat. The storage area should be accessible for maintenance and repairs.

Air Systems

Solar heating systems that circulate air through rooftop collectors commonly use a bin filled with rocks for heat storage (Fig. 7). The rock bin stores heat when hot air from the collectors is blown through the rocks. The air first enters a plenum at the top of the bin. It is distributed so that it flows through the bin evenly. As the air reaches the bottom of the bin, the rocks have absorbed most of its heat. The air enters another plenum at the bottom and is returned to the collectors for reheating.

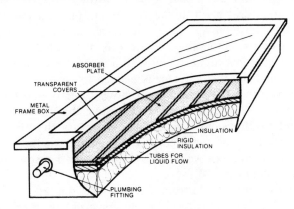

Fig. 6. A typical flat-plate solar collector.

227

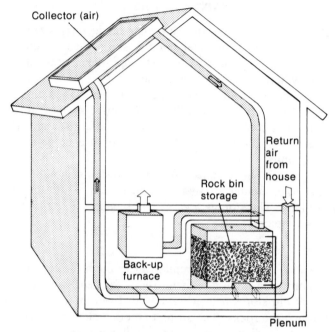

Fig. 7. An active system with rock bin storage.

To retrieve heat from the storage bin, house air is drawn into the lower plenum and blown upward through the rocks. The warmed air is drawn off at the top of the bin and distributed through the house. If the air is not warm enough, an auxiliary heater boosts its temperature before it is distributed.

It is not practical to store and withdraw heat simultaneously. If the hot air from the collector is needed to heat the house directly, it bypasses the rock bin and flows directly into the heat distribution system. Similarly, an air system that heats both the house and the household water should bypass the bin in the summer to heat the water directly. If a system heats the rock bin in order to preheat hot water, it will add to the house's cooling load in the summer.

Rock bins can be made from cinderblock, concrete, or wood. When treated plywood is used for bin construction, it should be lined with drywall and a vapor barrier. This will protect the rocks and the entire system from any gases released by the plywood. The greatest problem of air systems is air leakage from the storage bin, as well as from ducts and dampers. Because leaks drastically reduce the efficiency of the system, the bin must be tightly constructed and sealed. Sealing also prevents vermin and insects from entering.

Air should flow through the system at the rate of 2 to 5 cubic feet per minute for each square foot of collector. Ducts should be sized for a velocity of 5 to 10 feet per second, and the rock bin should provide ½ to 1 cubic foot of storage for every square foot of collector. This is roughly 2½ to 3 times the volume of a water tank that would provide an equivalent storage capacity in a liquid system.

The plenums at both the top and bottom of the bin ensure uniform airflow through the rock. Air filters are needed between both of the plenums and in the ducts to and from the bin. These filters prevent dust from blowing into either the collectors or the house. The filters should be inspected periodically and replaced when necessary.

Dense rock, such as river rock (which is predominantly quartz), performs best. The rocks should be of uniform size, roughly ¾ inch to 1½ inches in diameter. Before the rocks are put into the bin, they must be washed to remove dirt and insect eggs. Problems with mold, mildew, and insects can be prevented by keeping the rocks inside the bin dry.

Because of the weight of the rocks, the bin is usually located in the basement of the house. (Outside underground bins are another option, but they are difficult to insulate and waterproof and should be avoided. Ground water can ruin the insulation and may corrode the walls of the bin if the waterproofing materials fail.) To reduce heat loss, an indoor bin should be insulated to a value of not less than R-19 (R-30 or greater if it is outside). The ducts to and from the bin should be insulated to a value of R-16.

Liquid Systems

Water is the common storage medium in active systems that circulate liquid through collectors. From 1 to 2½ gallons of water are needed for every square foot of collector. The tanks used to store water can be made of concrete, steel, or fiberglass-reinforced plastic. All should be insulated to a value of R-19 or better, and all piping to a value of R-4. The cost and availability of suitable tanks vary widely across the country.

Steel tanks should be lined to prevent corrosion. Glass (domestic water heating only), epoxy, or butyl rubber (space heating only) can all be used as linings. When applying a liner to a tank, the manufacturer's recommendations should be followed carefully. It is especially important to consider the type of system, its volume and maximum temperature, and the wet vs. dry temperature limitations of the liner material. A rust inhibitor, such as disodium phosphate (Na_2HPO_4), can be used instead of a lining, but inhibitors degenerate and must be monitored regularly for pH balance. If the pH is not what is specified by the manufacturer, the inhibitor must be replaced. Concrete tanks, such as the ones used for septic systems, are sometimes used outdoors where steel is more likely to corrode. Since concrete is porous, the tanks should have either a vinyl or other liner to prevent leaks.

All tanks should have drains, as well as a means of access for routine maintenance, such as cleaning heat exchangers. Tanks should also be inspected for leaks before they are installed. And they should not be built into the structure of the house because they may someday need to be replaced. Temperature sensors for water storage tanks can and should be positioned for easy replacement.

If the local water is "hard" (i.e., has a high mineral content), a commercial water softener should be used to prevent problems with "scaling" (i.e., carbonate residue).

The water in the storage tank is heated in one of two ways. It is either circulated directly through the collectors (Fig. 8) or heated indirectly by a heat-transfer fluid that circulates from storage to the collectors within a "closed loop." In the latter instance, the transfer fluid absorbs heat from the collector and then passes through a heat exchanger, which transfers the heat to the water inside the storage tank. The heat exchanger is usually a coiled copper tube that conducts heat from the transfer fluid flowing through it to the storage water in which it is immersed. Because exposure to oxygen encourages corrosion, the heat exchanger should be placed completely below the water level inside the tank.

If the system is also providing hot water for domestic use, the heat exchanger needs to be double-walled so that drinking water cannot be contaminated by the heat-transfer fluid.

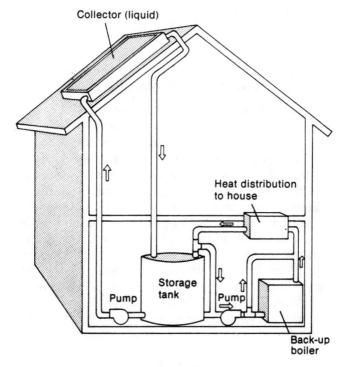

Fig. 8. Water tank storage.

HYBRID SYSTEMS

Elements of passive and active solar heating systems can be combined to create hybrid systems. The most common example uses a greenhouse collector (passive) linked to a rock storage bin (active) by way of fans and ducts. Hybrid systems offer more options to solar designers and builders and increase the overall versatility of solar heating technology. For example, collectors and/or storage in hybrid systems do not need to be located directly in the main living spaces; and interior walls, doors, and stairwells can be designed with less regard for free air movement

231

than is required with a strictly passive approach. It is important not to mistake a situation where two separate systems are present—one active, the other passive—as being hybrid. A system is considered hybrid only when components are mixed within a single system.

PHASE-CHANGE MATERIALS

Another storage medium, one that could be of use for both passive and active systems in the future, is phase-change materials. These are substances that store energy as latent heat. Latent heat is the heat absorbed by a material as it changes phase, such as from a solid to a liquid at constant temperature—for example, the melting of ice or wax. This heat is given up again as the liquid returns to its solid form. Two to five times more heat is stored per unit volume by this process than by raising the temperature of an equal volume of water, rock, or concrete. Researchers are investigating several types of materials, including salts and paraffin waxes. At present, storage in phase-change materials is more expensive than storage in water or rocks. Further testing and field applications will be necessary to prove the reliability of phase-change storage.

SUMMARY

Today's carpenters and other craftsmen must be conversant with solar energy because its installation in new or existing construction will many times require carpentry expertise.

There are two basic solar energy systems, active and passive, and a variety of storage systems. There is a variety of experimental systems extant, and hybrid systems which combine elements of the active and passive systems. A prime example of this is the greenhouse collector which is linked to a rock storage (active) bin by fans and ducts. Phase-change materials are also being investigated for their applicability in solar systems.

TABLE 1. Heat Capacities of Various Materials

Material	Specific Heat (BTU/lb/°F)	Density (lb/ft³)	Heat Capacity (BTU/ft³/°F)
Water	1.00	62.5	62.5
Concrete	0.27	140	38
Brick	0.20	140	28
Gypsum	0.26	78	20.3
Marble	0.21	180	38
Asphalt	0.22	132	29.0
Sand	0.191	94.6	18.1
White Pine	0.67	27	18.1
Air (75°F)	0.24	0.075	0.018

REVIEW QUESTIONS

1. What is the difference between the active and passive solar systems?
2. What is direct and indirect heat gain?
3. Explain one hybrid system.
4. What are phase-change materials?

CHAPTER 16

Paneling

A major development in wall covering over the last twenty-five years is paneling. It can take walls in any condition, from ones that are slightly damaged to ones that are virtually decimated, and give them a sparkling new face. Paneling is a key improvement material for the carpenter and do-it-yourselfer.

Paneling is available in an almost limitless variety of styles, colors, and textures. Prefinished paneling may be plywood or processed wood fiber products. Paneling faces may be real hardwood or softwood veneers finished to enhance their natural texture, grain, and color. Other faces may be printed, paper overlaid, or treated to simulate wood grain. Finishing techniques on both real wood and simulated wood-grain surfaces provide a durable and easily maintained wall. Cleaning usually consists of wiping with a damp cloth.

HARDWOOD AND SOFTWOOD PLYWOOD

Plywood panelings are manufactured with a face, core, and back veneer of softwood, hardwood, or both (Fig. 2). The face and back veneer wood grains run vertically, the core horizontally. This lends strength and stability to the plywood. Hardwood and softwood plywood wall paneling is normally ¼ inch thick with some paneling ranging up to $7/16$ inch thick.

The most elegant and expensive plywood paneling has the real hardwood or softwood face veneers—walnut, birch, elm, oak, cherry, cedar, pine, fir. Other plywood paneling may have a veneer of tropical hardwood. Finishing techniques include em-

Fig. 1. As a wall covering, paneling is hard to beat. Courtesy of Georgia-Pacific.

Fig. 2. Plywood paneling is composed of a back, core, and face.

bossing, antiquing, or color toning to achieve a wood grain or decorative look. Panels may also have a paper overlay with wood grain or patterned paper laminated to the face. The panel is then grooved and finished. Most of these panelings are $^5/_{32}$ inch thick.

WOOD FIBER PANELING

Processed wood fiber—particleboard or hardboard—wall panelings are also available with grain-printed paper overlays or printed surfaces. These prefinished panels are economical yet attractive. Thicknesses available are $^5/_{32}$, $^3/_{16}$, and ¼ inch.

Wood fiber paneling requires special installation techniques. Look for the manufacturer's installation instructions printed on the back of each panel.

Other paneling includes hardboard and hardboard with a tough plastic coating that makes the material suitable for use in high-moisture areas, such as the bathroom.

GROOVE TREATMENT

Most vertical wall paneling is "random grooved" with grooves falling on 16-inch centers so that nailing over studs will be consistent. A typical random-groove pattern may look like the one illustrated in Fig. 3.

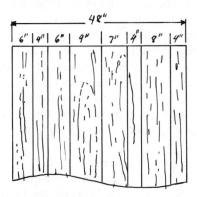

Fig. 3. Typical grooves on a panel.

Other groove treatments include uniform spacing (4, 8, 12, or 16 inch) and cross-scored grooves randomly spaced to give a "planked" effect. Grooves are generally striped darker than the panel surface. They are cut or embossed into the panel in V-grooves or channel grooves. Less expensive paneling sometimes has a groove "striped" on the surface.

BUYING TIPS

Like most products for the home, paneling is available in a wide range of prices, from as little as $5 per panel to over $30 per panel retail. Contractors and carpenters get a discount (usually 10 percent).

Generally, paneling with a simulated wood-grain finish is less expensive than real wood-surfaced panels. Printed or plywood overlaid paneling is available for about $9–$15 per panel. Paneling with wood fiber substrate costs $5–$8 per sheet.

How To Figure Amount of Material Needed

There are several ways to estimate the amount of paneling you will need. One way is to make a plan of the room. Start your plan on graph paper that has ¼-inch squares. Measure the room width, making note of window and door openings. Translate the measurements to the graph paper. Example: Let us assume that you have a room 14′ × 16′ in size. If you let each ¼-inch square represent 6 inches, the scale is 1 inch equals 2 feet, so you end up with a 7″ × 8″ rectangle. Indicate window positions, doors (and the direction in which they open), a fireplace, and other structural elements. Now begin figuring your paneling requirements. Total all four wall measurements: 14′ + 16′ + 14′ + 16′ = 60′. Divide the total perimeter footage by the 4-foot panel width: 60′ ÷ 4′ = 15 panels. To allow for door, window, and fireplace cutouts, use the deductions as follows (approximate): door—½ panel (A), window—¼ panel (B), fireplace—½ panel (C) (Fig. 4).

To estimate the paneling needed, deduct the cutout panels from your original figure: 15 panels − 2 panels = 13 panels. If the perimeter of the room falls between the figures in Table 1, use the next higher number to determine panels required. These

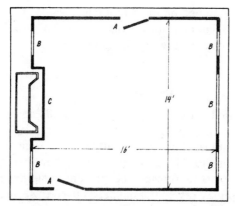

Fig. 4. *A way to calculate panel needs with graph paper. See text for details.*

figures are for rooms with 8-foot wall heights or less. For higher walls, add in the additional materials needed above the 8-foot level.

New paneling should be stored in a dry location. In new construction, freshly plastered walls must be allowed to dry thoroughly before panel installation. Prefinished paneling is moisture-resistant, but like all wood products, it is not waterproof and should not be stored or installed in areas subject to excessive moisture. Ideally the paneling should be stacked flat on the floor with sticks between sheets (Fig. 5) to allow air circulation,

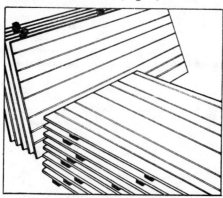

Fig. 5. *Before installation, store panels with sticks between. This allows air to circulate.*

239

or it should be propped on the 8-foot edge. Panels should remain in the room 48 hours prior to installation to permit them to acclimatize to temperature and humidity.

INSTALLATION

Installing paneling over existing straight walls above grade requires no preliminary preparation.

First, locate studs. Studs are usually spaced 16 inches on center, but variations of 24-inch centers and other spacing may be found. If you plan to replace the present molding, remove it carefully—you may reveal the nailheads used to secure the plaster lath or drywall to the studs. These nails mark stud locations. If you cannot locate the studs, start probing into the wall surface to be paneled with a nail or a small drill until you hit solid wood. Start 16 inches from one corner of the room with your first try, making test holes at ¾-inch intervals on each side of the initial hole until you locate the stud (Figs. 6 and 7).

Studs are not always straight, and so it is a good idea to probe at several heights. Once a stud is located, make a light pencil mark at floor and ceiling to position all studs, then snap a chalk line at 4-foot intervals (the standard panel width).

TABLE 1

Perimeter	No. of 4' × 8' panels needed (without deductions)
36'	9
40'	10
44'	11
48'	12
52'	13
56'	14
60'	15
64'	16
68'	17
72'	18
92'	23

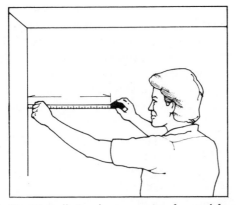

Fig. 6. Start installation by measuring for stud from corner.

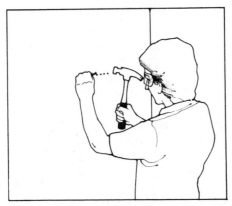

Fig. 7. Drive nail into wall to determine position of stud. Mark it and others with chalk line.

Measuring and Cutting

Start in one corner of the room and measure the floor-to-ceiling height for the first panel. Subtract ½ inch to allow for clearance top and bottom. Transfer the measurements to the first panel and mark the dimensions in pencil, using a straightedge for a clean line. Use a sharp crosscut handsaw with ten or more teeth to the inch, or a plywood blade in a table saw, for all cutting. Cut with the panel face up. If you are using a portable circular saw or a saber saw, mark and cut panels from the back.

241

Cutouts for door and window sections, electrical switches, and outlets or heat registers require careful measurements. Take your dimensions from the edge of the last applied panel for width, and from the floor and ceiling line for height. Transfer the measurements to the panel, checking as you go. Unless you plan to add molding, door and window cutouts should fit against the surrounding casing. If possible, cutout panels should meet close to the middle over doors and windows.

An even easier way to make cutouts is with a router. You can tack a panel over an opening—a window, say—after removing the trim. Then use the router to trim out the waste paneling in the opening, using the edge of the opening as a guide for the router bit to ride on. Then you can tack the molding or trim back in place.

For electrical boxes, shut off the power and unscrew the protective plate to expose the box. Then paint or run chalk around the box edges and carefully position the panel. Press the paneling firmly over the box area, transferring the outline to the back of the panel (Fig. 8). To replace the switchplate, a ¼-inch spacer or washer may be needed between the box screw hole and the switch or receptacle.

Drive small nails in each corner through the panel until they protrude through the face. Turn the panel over, drill two ¾-inch holes just inside the corners, and use a keyhole or saber saw to

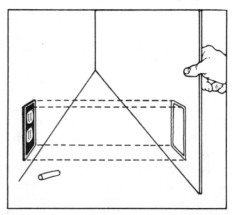

Fig. 8. One way to make a cutout is to smear lipstick on plate, press panel against it, then pull it back. Outline will be on it.

make the cutout (Fig. 9). The hole can be up to ¼ inch over size and still be covered when the protective switchplate is replaced.

Securing Panels

Put the first panel in place and butt to an adjacent wall in the corner. Make sure it is plumb and that both left and right panel edges fall on solid stud backing. Most corners are not perfectly true, however, so you will probably need to trim the panel to fit into the corner properly. Fit the panel into the corner, checking with a level to be sure the panel is plumb vertically. Draw a mark along the panel edge, parallel to the corner. On rough walls like masonry, or on walls adjoining a fireplace, scribe or mark the panel with a compass on the inner panel edge, then cut on the scribe line to fit (Fig. 10). Scribing and cutting the inner panel edge may also be necessary if the outer edge of the panel does not fit directly on a stud. The outer edge must fall on the center of a stud to allow room for nailing your next panel. Before installing the paneling, paint a stripe of color to match the paneling groove color on the wall location where panels meet. The gap between panels will not show.

Most grooved panels are random grooved to create a solid lumber effect, but there is usually a groove located every 16 inches. This allows most nails to be placed in the grooves, falling directly on the 16-inch stud spacing. Regular smallheaded finish nails or colored paneling nails can be used. For paneling directly

Fig. 9. Use a keyhole saw for cutout.

243

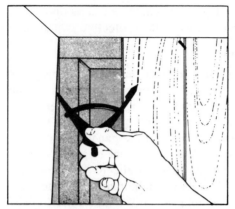

Fig. 10. Method of making uneven edge on panel.

onto studs, 3d (1¼ inch) nails are recommended; but if you must penetrate backerboard, plaster, or drywall, 6d (2 inch) nails are needed to get a solid bite in the stud. Space nails 6 inches apart on the panel edges and 12 inches apart in the panel field (Fig. 11). Nails should be countersunk slightly below the panel surface with a nail set, then hidden with a matching putty stick. Colored nails eliminate the need to do this. Use 1-inch colored nails to apply paneling to studs, 1⅝-inch nails to apply paneling through plasterboard or plaster.

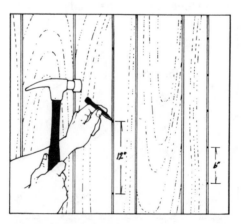

Fig. 11. Nails should be driven as indicated.

Adhesive Installation

Using adhesive to install paneling eliminates countersinking and hiding nailheads. Adhesive may be used to apply paneling directly to studs or over existing walls as long as the surface is sound and clean.

Paneling must be cut and fit prior to installation. Make sure the panels and walls are clean—free from dirt and particles—before you start. Once applied, the adhesive makes adjustments difficult.

A calking gun with adhesive tube is the simplest method of application. Trim the tube end so that a ⅛-inch-wide adhesive bead can be squeezed out. Once the paneling is fitted, apply beads of adhesive in a continuous strip along the top, bottom, and both ends of the panel. On intermediate studs, apply beads of adhesive 3 inches long and 6 inches apart (Fig. 12).

With scrap plywood or shingles used as a spacer at the floor level, set the panel in place and press firmly along the stud lines, spreading the adhesive on the wall. Using a hammer with a padded wooden block or a rubber mallet, tap over the glue lines to assure a sound bond between panel and backing.

Some adhesives require panels to be placed against the adhesive, then gently pulled away and allowed to stand for a few minutes for the solvent to "flash off" and the adhesive to set up. The panel is then repositioned and tapped home.

Fig. 12. Adhesive can also be used for paneling.

Uneven Surfaces

Most paneling installations require no preliminary preparations. When you start with a sound level surface, paneling is quick and easy.

Not every wall is a perfect wall, however. Walls can have chipped, broken, and crumbly plaster, peeling wallpaper, or gypsum board punctured by a swinging doorknob or a runaway tricycle on a rainy afternoon. Or walls can be of rough poured concrete or cinderblock. Walls must be fixed before you attempt to cover them with paneling.

Most problem walls fall into one of two categories. Either you are dealing with plaster or gypsum board applied to a conventional wood stud wall, or you are facing an uneven masonry wall—brick, stone, cement, or cinderblock, for example. The solution to both problems is the same, but getting there calls for slightly different approaches.

On conventional walls, clean off any obviously damaged areas. Remove torn wallpaper, scrape off flaking plaster and any broken gypsum board sections. On masonry walls, chip off any protruding mortar. Don't bother making repairs; there is an easier solution.

The paneling will be attached to furring strips. Furring strips are either 1″ × 2″ lumber or ⅜ or ½-inch plywood strips cut 1½ inches wide. The furring strips are applied horizontally 16 inches apart on center on the wall (based on 8-foot ceiling) with vertical members at 48-inch centers where the panels butt together.

Begin by locating the high spot on the wall. To determine this, drop a plumb line (Fig. 13). Fasten your first furring strip, making sure that the thickness of the furring strips compensates for the protrusion of the wall surface. Check with a level to make sure that each furring strip is flush with the first strip (Fig. 14). Use wood shingles or wedges between the wall and strips to assure a uniformly flat surface (Fig. 15). The furred wall should have a ½-inch space at the floor and ceiling with the horizontal strips 16 inches apart on center and the vertical strips 48 inches apart on center (Fig. 16). Remember to fur around doors, windows, and other openings (Fig. 17).

On stud walls, the furring strips can be nailed directly through

Fig. 13. A plumb bob can be used to determine high spots in wall.

Fig. 14. Check furring strips for trueness with level.

the shims and gypsum board or plaster into the studs. Depending on the thickness of the furring and wall covering, you'll need 6d (2 inch) or 8d (2½ inch) common nails.

Masonry walls are a little tougher to handle. Specially hardened masonry nails can be used, or you can drill a hole with a carbide tipped bit, insert wood plugs or expansion shields, and nail or bolt the shimmed frame into place (Fig. 18).

If the masonry wall is badly damaged, construct a 2″ × 3″ stud wall to install paneling on (Fig. 19). Panels may be directly installed on this wall (Fig. 20).

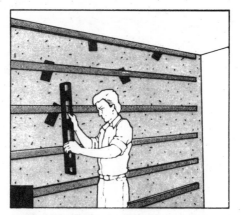

Fig. 15. Wedges will bring strips true.

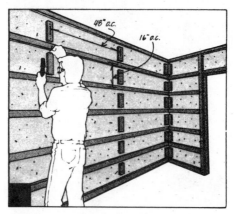

Fig. 16. Spacing of strips is as shown.

DAMP WALLS

Masonry walls, besides being uneven, often present a more difficult problem—dampness. Usually, damp walls are found partially or fully below ground level. Damp basement walls may result from two conditions: (1) seepage of water from outside walls; or (2) condensation, moist warm air within the home that condenses or beads up when it comes in contact with cooler outside walls. Whatever the cause, the moisture must be eliminated before you consider paneling.

Fig. 17. Don't forget to fur around windows and doors.

Seepage may be caused by leaky gutters, improper grade, or cracks in the foundation. Any holes or cracks should be repaired with concrete patching compound or with special hydraulic cement designed to plug active leaks.

Weeping or porous walls can often be corrected with an application of masonry waterproofing paints. Formulas for wet and dry walls are available, but the paint must be scrubbed onto the surface to penetrate the pores, hairline cracks, and crevices.

Condensation problems require a different attack. Here, the answer is to dry out the basement air. The basement should be provided with heat in the winter and cross-ventilation in the summer. Wrap cold-water pipes with insulation to reduce sweating. Vent basement washer, dryer, and shower directly to the outside. It may pay big dividends to consider installing a dehumidifying system to control condensation.

When you have the dampness problem under control, install plumb, level furring strips on the wall. Apply insulation if desired. Then line the walls with a polyvinyl vapor/moisture barrier film installed over the furring strips (Fig. 21). Rolls of this inexpensive material can be obtained and installed, but be sure to provide at least a 6-inch lap where sections meet.

Apply paneling in the conventional method. If the dampness is so serious that it cannot be corrected effectively, then other steps must be taken. Prefinished hardwood paneling is manufac-

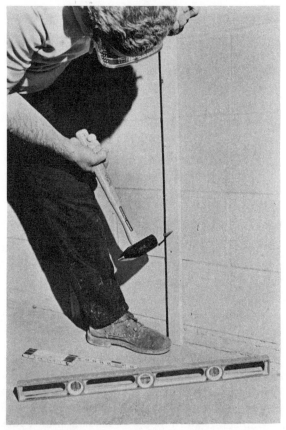

**Fig. 18. Cut nails and a heavy hammer can be used to secure furring.
Courtesy of Vaughn & Bushnell.**

tured with interior glue and must be installed in a dry setting for
satisfactory performance.

PROBLEM CONSTRUCTION

Sometimes the problem is out-of-square walls, uneven floors,
or studs placed out of sequence. Or trimming around a stone or
brick fireplace may be required, or making cutouts for wall
pipes, or handling beams and columns in a basement.

Any wall to be paneled should be checked for trueness. If the

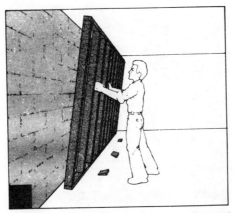

Fig. 19. If wall is very rough, a 2 × 3 wall can be built.

Fig. 20. Panels should be lined up around it in best match.

wall is badly out-of-plumb, it must be corrected before you install paneling. Furring strips should be applied so that they run at right angles to the direction of the panel application. Furring strips of ⅜" × 1½" plywood strips, or 1" × 2" lumber should be used. Solid backing is required along all four panel edges on each panel. Add strips wherever needed to ensure this support. It is a good idea to place the bottom furring strip ½ inch from the floor. Leave a ¼-inch space between the horizontal strips and the vertical strips to allow for ventilation.

If the plaster, masonry, or other type of wall is so uneven that

251

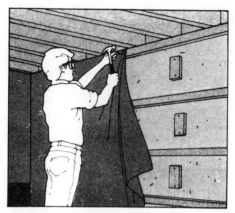

Fig. 21. If needed, polyethylene makes a good moisture barrier.

it cannot be trued by using furring strips and shimming them out where wall bellows, 2″ × 3″ studs may be necessary. You can use studs flat against the wall to conserve space. Use studs for top and bottom plates, and space vertical studs 16 inches on center. Apply paneling direct to studs or over gypsum or plywood backerboard.

Out-of-sequence studs (i.e., studs not on regular centers) require a little planning, but usually they are not a serious problem. Probe to find the exact stud locations as described earlier, snap chalk lines, and examine the situation. Usually you will find a normal stud sequence starting at one corner, then, perhaps where the carpenter framed for a mid-wall doorway, the spacing abruptly changes to a new pattern. Start paneling in the normal manner, using panel adhesive and nails. When you reach the changeover point, cut a filler strip of paneling to bridge the odd stud spacing and then pick up the new pattern with full-size panels.

Occasionally you will find a stud or two applied slanted. The combination of panel adhesive plus the holding power of the ceiling and baseboard moldings generally solve this problem.

Uneven floors can usually be handled with shoe molding. Shoe, ½″ × ¾″, is flexible enough to conform to moderate deviations. If the gap is greater than ¾ inch, the base molding or panel should be scribed with a compass to conform to the floor line. Spread the points slightly greater than the gap, hold the compass vertically, and draw the point along the floor, scribing a pencil line on the base molding. Remove the molding trim to the new line with a coping or saber saw and replace. If you have a real washboard of a floor, then you have a floor problem, not a wall problem. Either renail the floor flat or cover it with a plywood or particleboard underlayer before paneling.

The compass trick is also used to scribe a line where paneling butts into a stone or brick fireplace (Fig. 22). Tack the panel in place temporarily, scribe a line parallel to the fireplace edge, and trim. Check to be sure that the opposite panel edge falls on a stud before applying.

Fig. 22. Scribe may be used to make panel edge fit when wall is rough or masonry.

Older homes may have heating pipes, usually pipes near the ceiling, protruding through the wall. To panel around pipes or heating ducts, make a double cut the width of the hole diameter, install panel, and nail the cutout section above the pipe. Colored putty sticks will blend cut edges, and metal collars are available to snap around the pipes for a clean job (Fig. 23).

Vertical heating pipes in corners, overhead beams, or support columns in a basement are unattractive elements that are best hidden from sight. Construct a simple 1″ × 2″ lumber framework, tying it to the adjacent wall or ceiling. Apply your paneling and cover the joints with outside corner molding for a neat job (Fig. 24).

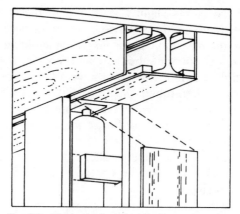

Fig. 23. Method of enclosing beams or pipes.

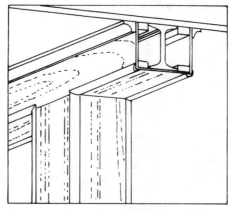

Fig. 24. Beams are hidden very well.

SUMMARY

Paneling is one of the most popular ways to dress up walls, and is a job which can be readily done by do-it-yourselfers as well as carpenters.

There is a variety of paneling available, ranging from plywood, hardboard, and plastic coated materials. Some paneling has simulated fine-wood finishes while other types (plywood) have a veneer of fine wood over a core.

Installing paneling is usually relatively easy, but in some cases walls will be so uneven as to require a plumb wall of furring strips or studs be installed to be a base for the material. In some cases, also, dampness may be a problem and this must be corrected before the paneling is installed.

REVIEW QUESTIONS

1. Briefly describe the types of paneling.
2. How is paneling cut?
3. How are furring strips secured to masonry?

Index